M000011433
SUMMIT MATH
Learn at your OWN pace.
ALGEBRA 2
second edition
3
QUADRATIC EQUATIONS & PARABOLAS

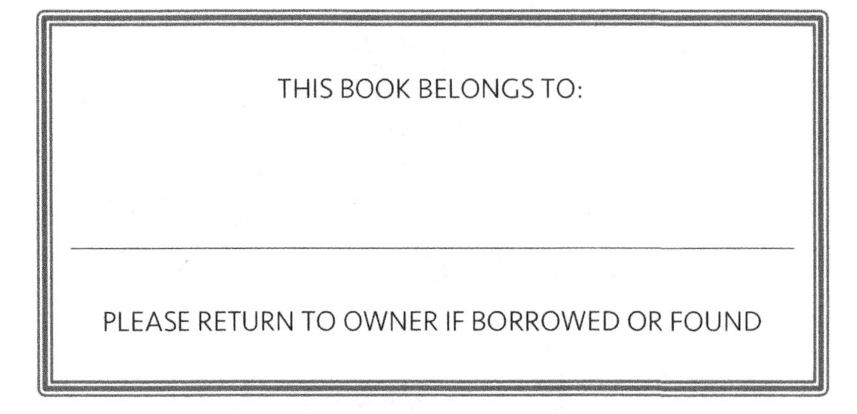

DEDICATION

To Lauren, Chloe, Dawson and Teagan

ACKNOWLEDGEMENTS

I started writing these books in 2013 to help my students learn better. I kept writing them because I received encouraging feedback from students, parents and teachers. Thank you to all who have used these books, pointed out my mistakes, and made suggestions along the way. Thank you to all of the students and parents who asked me to keep writing more books. Thank you to my family for supporting me through every step of this journey.

This book was typeset in the following fonts:
Seravek + Mohave + *Heading Pro*

Graphics in Summit Math books are made using the following resources:
Microsoft Excel | Microsoft Word | Desmos | Geogebra | Adobe Illustrator

First printed in 2017

Printed in the U.S.A.

Summit Math Books are written by Alex Joujan.

www.summitmathbooks.com

Learning math through Guided Discovery:

A Guided Discovery learning experience is designed to help you experience a feeling of discovery as you learn each new topic.

Why this curriculum series is named Summit Math:

Learning through Guided Discovery can be compared to climbing a mountain. Climbing and learning both require effort and persistence. In both activities, people naturally move at different paces, but they can reach the summit if they keep moving forward. Whether you race rapidly through these books or step slowly through each scenario, this curriculum is designed to keep advancing your learning until you reach the end of the book.

Guided Discovery Scenarios:

The Guided Discovery Scenarios in this book are written and arranged to show you that new math concepts are related to previous concepts you have already learned. Try to fully understand each scenario before moving on to the next one. To do this, try the scenario on your own first, check your answer when you finish, and then fix any mistakes, if needed. Making mistakes and struggling are essential parts of the learning process.

Homework and Extra Practice Scenarios:

After you complete the scenarios in each Guided Discovery section, you may think you know those topics well, but over time, you will forget what you have learned. Extra practice will help you develop better retention of each topic. Use the Homework and Extra Practice Scenarios to improve your understanding and to increase your ability to retain what you have learned.

The Answer Key:

The Answer Key is included to promote learning. When you finish a scenario, you can get immediate feedback. When the Answer Key is not enough to help you fully understand a scenario, you should try to get additional guidance from another student or a teacher.

Star symbols:

Scenarios marked with a star symbol ★ can be used to provide you with additional challenges. Star scenarios are like detours on a hiking trail. They take more time, but you may enjoy the experience. If you skip scenarios marked with a star, you will still learn the core concepts of the book.

To learn more about Summit Math and to see more resources:

Visit www.summitmathbooks.com.

As you complete scenarios in this part of the book, follow the steps below.

Step 1: Try the scenario.
Read through the scenario on your own or with other classmates. Examine the information carefully. Try to use what you already know to complete the scenario. Be willing to struggle.

Step 2: Check the Answer Key.
When you look at the Answer Key, it will help you see if you fully understand the math concepts involved in that scenario. It may teach you something new. It may show you that you need guidance from someone else.

Step 3: Fix your mistakes, if needed.
If there is something in the scenario that you do not fully understand, do something to help you understand it better. Go back through your work and try to find and fix your errors. Mistakes provide an opportunity to learn. If you need extra guidance, get help from another student or a teacher.

After Step 3, go to the next scenario and repeat this 3-step cycle.

NEED EXTRA HELP?
watch videos online

Teaching videos for every scenario in the Guided Discovery section of this book are available at www.summitmathbooks.com/algebra-2-videos.

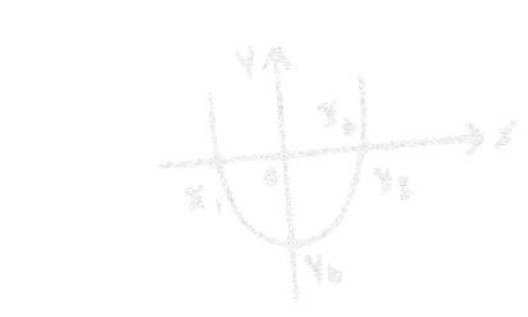

CONTENTS

Section 1

INTRODUCTION TO QUADRATIC FUNCTIONS

1. A baseball game is tied and there are already two outs in the bottom of the 9th inning. The pitcher winds up and throws the ball, hoping it goes by the batter for a strike. The batter swings the bat and sends the ball sailing through the air toward the outfield fence. The ball follows a smooth, arching path, modeled by the equation $H=-0.01x^2+3x+3$, where H is the height of the ball after it has traveled x feet horizontally. If the outfield fence is 350 feet away and 9 feet high, will the ball clear the fence?

2. In the previous scenario, if nothing stops the ball or changes the direction of its path, how far will it travel horizontally before it lands on the ground, rounded to the nearest foot? Write an equation to answer this question, but do not solve the equation.

When the ball hits the ground, it's height is 0 feet. Thus, one way to find the distance the ball travels is to try out different x-values to see if they produce a height of 0. You could let x be 350, or 351, or 352, or..., but this guessing might take awhile. Another way to find this distance is to let the equation tell you the x-value for which the ball has a height of 0. This would lead you to consider the equation $0=-0.01x^2+3x+3$.

At this point, you might know of only one way to solve this type of equation: factor the quadratic expression and set each of the factors equal to 0. Unfortunately, this equation cannot be solved using the factoring method, so you will need to find another way to solve the equation. In upcoming scenarios, you will learn how to find solutions to any equation that has a quadratic structure, which will especially help when factoring is not an option.

Use this page to record important ideas in the previous section or
for any other writing that helps you learn the topics in this book.

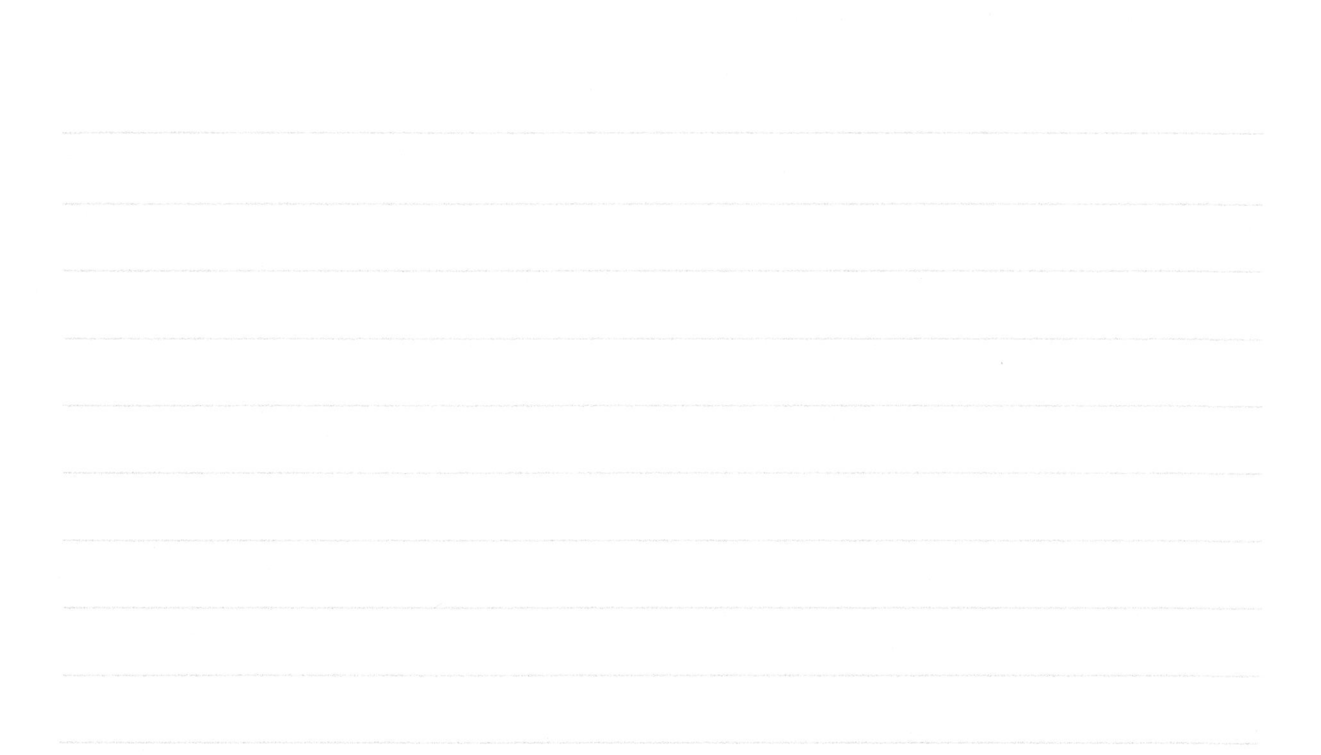

Section 2
FACTORING REVIEW

Let's start by making sure you are familiar with the topic of factoring.

3. Fill in the blank to make the factors match the expression written below.

a. $(x+\_\_)(x-3)$
$x^2+8x-33$

b. $(2x-\_\_)(x-5)$
$2x^2-\_\_\_+15$

★c. $x(x+\_\_)(x-\_\_)$
x^3+2x^2-35x

4. Fill in the blank to make each expression match the factors written below.

a. $x^2+\_\_\_x-21$
$(x+7)(x-3)$

b. $x^2-10x+25$
$(x-\_\_)(x-5)$

★c. $2x^2+\_\_\_x-12$
$2(x-6)(x+1)$

5. Factor each of the following expressions.

a. $x^2-8x+12$ b. x^2-36 ★c. $x^2-14x+49$

6. Factor each of the following expressions.

a. $2x^2+7x+6$ b. $6x^2-x-35$ ★c. $9x^2+12x+4$

7. Solve each equation.

a. $x^2+10x+9=0$ b. $x^2-49=0$ ★c. $4x^2-20x+25=0$

8. You are given a new function and you cannot see its graph. How can you find the coordinates of the x-intercepts of the function?

9. Where does the function $f(x)=x^2+3x-10$ cross the x-axis?

Section 3
REVIEW RADICAL EXPRESSIONS

As you learn how to solve quadratic equations using methods other than factoring, you will need to be familiar with radical expressions, so it will be helpful to review them now.

10. Simplify each expression.

a. $\sqrt{9}$ b. $\sqrt{49}$ c. $\sqrt{121}$

11. Simply each expression by separating the integer into two factors, where one of the factors is a perfect square. For example, $\sqrt{32}$ can be written as $\sqrt{16}\sqrt{2}$, or $4\sqrt{2}$.

a. $\sqrt{27}$ b. $\sqrt{24}$ c. $\sqrt{80}$

12. Now simplify these expressions that involve fractions.

a. $\sqrt{\frac{4}{9}}$ b. $\sqrt{\frac{5}{9}}$ c. $\sqrt{\frac{8}{9}}$ d. $\sqrt{\frac{9}{9}}$ ★e. $\sqrt{\frac{27}{9}}$

13. In a previous lesson, you learned about radical expressions like $\sqrt{2}$ and $-\sqrt{3}$. You also learned that an expression like $\frac{3}{\sqrt{5}}$ can be changed to make the denominator rational. This is done by multiplying the fraction by a disguised form of one. Complete the steps shown below by filling in missing parts.

$$\frac{3}{\sqrt{5}} \cdot \frac{}{} = \frac{}{5}$$

14. Change each fraction below to make its denominator a rational number.

a. $\frac{2}{\sqrt{3}}$ b. $\frac{10}{\sqrt{2}}$ ★c. $\frac{7\sqrt{3}}{\sqrt{7}}$

15. Simplify the square root of each fraction below.

a. $\sqrt{\frac{16}{25}}$ b. $\sqrt{\frac{3}{4}}$ c. $\sqrt{\frac{4}{3}}$

Section 4
IMAGINARY NUMBERS

16. How would you simplify $\sqrt{-1}$?

The value of $\sqrt{1}$ is 1 because $1^2 = 1$. When you consider the value of $\sqrt{-1}$, the logical result is that $\sqrt{-1}$ has no value because no number can be squared to equal -1. Many calculators will display an error message when you ask the calculator to give you the value of $\sqrt{-1}$. However, mathematicians have addressed this issue by setting aside a specific name for $\sqrt{-1}$.

17. Read the following statement as well as you can.

    I do not completely understand th$\sqrt{-1}$s yet, but I th$\sqrt{-1}$nk I eventually w$\sqrt{-1}$ll. I am confused about th$\sqrt{-1}$s new square root, though, because $\sqrt{-1}$f the square root of "−1" $\sqrt{-1}$s not a real number, why do mathemat$\sqrt{-1}$c$\sqrt{-1}$ans give $\sqrt{-1}$t a name?

18. What expression do you think mathematicians use to represent $\sqrt{-1}$?

19. What is the value of $\sqrt{-4}$?

20. What is the value of $\sqrt{-9}$?

21. Find the value of A that makes each statement true.

    a. $\sqrt{-16} = Ai$ b. $\sqrt{-49} = Ai$ c. $\sqrt{A} = 9i$ ★d. $\sqrt{-\frac{4}{9}} = Ai$ ★e. $\sqrt{A} = \frac{1}{2}i$

22. Since $\sqrt{12}$ is equivalent to $2\sqrt{3}$, it follows that $\sqrt{-12}$ can be written as $2\sqrt{3}i$. However, in this case, it is typical to write $2i\sqrt{3}$ instead. Why is $2i\sqrt{3}$ the preferred way to express $\sqrt{-12}$?

23. Simplify each expression.

    a. $\sqrt{-8}$ b. $\sqrt{-20}$ ★c. $\sqrt{-44}$

24. How would you write these expressions in simplified form?

a. $\sqrt{-\frac{16}{25}}$ b. $\sqrt{-\frac{3}{4}}$ ★c. $\sqrt{-\frac{1}{2}}$

You now see that i represents $\sqrt{-1}$, but you may not know why the letter i is chosen for this purpose. The numbers on a typical number line are called the real numbers. Expressions like $\sqrt{-2}$, $\sqrt{-3}$, $\sqrt{-4}$, ... etc. do not have real values, so they have no location on the real number line. Thus, they are called imaginary numbers. These imaginary numbers can be written as $i\sqrt{2}$, $i\sqrt{3}$, $2i$, … . Since i is part of each of these numbers, i is known as the imaginary number.

25. Since the value of $\sqrt{-1}$ is i, what is the value of i^2 ?

26. To see why $i^2 = -1$, consider what you already know about square roots. The value of $\sqrt{4}$ is 2 because 2^2 is 4. These are two defined amounts that go together. Continue this logical review below.

a. The value of $\sqrt{9}$ is 3 because $(\_\_\_\_)^2 = \_\_\_\_$.

b. $\sqrt{49} = 7$ because $(\_\_\_\_)^2 = \_\_\_\_$.

c. Since $\sqrt{24} = 2\sqrt{6}$, it follows that $(\_\_\_\_)^2 = \_\_\_\_$.

d. Since $\sqrt{\frac{4}{9}} = \_\_\_\_$, it follows that $\left(\quad \right)^2 = \_\_\_\_$.

e. And finally, since $\sqrt{-1} = \_\_\_\_$, it follows that $(\quad)^2 = \_\_\_\_$.

27. Simplify each expression below.

a. $(2i)^2$ b. $(-2i)^2$ c. $(-5i)^2$ ★d. $(i\sqrt{3})^2$

Use this page to record important ideas in the previous section or for any other writing that helps you learn the topics in this book.

Section 5
QUADRATIC EQUATIONS

28. Solve the following equations. It will help to remember that an equation is considered to be solved when you have identified all possible values of x that make the equation true.

a. $x^2 = 25$ b. $2x^2 = 18$ ★c. $5x^2 - 1 = 79$

29. If you work too quickly through the previous equations, you might think they each have one solution, but they actually each have two solutions. Why do the previous equations each have two solutions?

30. If $M^2 = 9$, then M can be 3 or -3. There are 2 numbers that make $M^2 = 9$.

a. If $(x+1)^2 = 9$, then $x+1 =$ ______. b. If $(x-7)^2 = 25$, then $x-7 =$ ______.

31. Now consider the equation $(x-4)^2 = 3$. One way to solve this equation is to square the binomial first and then use factoring. If you tried this strategy, your work might follow the steps below.

$$(x-4)^2 = 3 \rightarrow x^2 - 8x + 16 = 3 \rightarrow x^2 - 8x + 13 = 0 \rightarrow (\underline{\quad ? \quad})(\underline{\quad ? \quad}) = 0$$

The trinomial $x^2 - 8x + 13$ cannot be factored, but before you assume this equation cannot be solved, look at the original equation below and try a different approach. Start by repeating the first step in the previous scenario and then try to proceed from there.

$(x-4)^2 = 3$

32. In the previous scenario, the equation $(x-4)^2 = 3$ has two solutions: $4+\sqrt{3}$ and $4-\sqrt{3}$. To see these numbers in a different form, input each expression into a calculator and write its decimal value. Round to the nearest tenth.

33. Solve the equation below, which has a similar structure to the equation in the previous scenario.

$(x-1)^2 = 25$

34. Solve each equation.

a. $(x+11)^2=81$

b. $(x-7)^2=2$

35. The equation below looks like the previous ones, with a slight difference. Describe the first operation you can perform to make the equation look like the ones in the previous scenario.

$2(x+4)^2=18$

36. Solve the equation in the previous scenario.

37. Solve each equation.

a. $8(x+11)^2=40$

b. $3(x-5)^2=6$

38. Some equations do not have real solutions. Use what you learned in previous scenarios to solve the following equations.

a. $x^2=-9$

b. $x^2=-5$

39. There are two solutions to each equation in the previous scenario, but it may be difficult to see why this occurs. To help you see this, substitute each solution into the original equation to see that it makes the equation true.

Use this page to record important ideas in the previous section or for any other writing that helps you learn the topics in this book.

Section 6

SOLVING QUADRATIC EQUATIONS BY COMPLETING THE SQUARE

You are almost ready to learn a new method for solving quadratic equations, but you first need to review how to square a binomial. For example, $(x-3)^2$ or $(x-3)(x-3)$ is equivalent to the expression x^2-6x+9. There is a distinct relationship between the numbers in these two expressions and you will analyze this relationship in the next scenario.

40. Consider the following pairs of equivalent expressions. Fill in the blanks with values that make each pair of expressions equal.

a. $(x+1)^2 = x^2 + \_\_\_ x+1$ b. $(x-2)^2 = x^2 - \_\_\_ x + \_\_\_$

41. Fill in the blanks with values that make each pair of expressions equal.

a. $(x+3)^2 = x^2+6x+\_\_\_$ b. $(x+\_\_\_)^2 = x^2+8x+\_\_\_$ c. $(x-\_\_\_)^2 = x^2+\_\_\_ x+49$

42. Fill in the blanks with values that make each pair of expressions equal.

a. $(x+\_\_\_)^2 = x^2+2x+\_\_\_$ b. $(x-\_\_\_)^2 = x^2-4x+\_\_\_$ c. $(x-\_\_\_)^2 = x^2-16x+\_\_\_$

43. Suppose the expression $(x+A)^2$ is equivalent to the expression x^2+Bx+C.

a. What is the relationship between *A* and *B*?

b. What is the relationship between *A* and *C*?

c. What is the relationship between *B* and *C*?

44. Fill in each blank with a number that makes each expression a perfect square trinomial.

a. $x^2+12x+\_\_\_$ b. $x^2-2x+\_\_\_$ c. $x^2-16x+\_\_\_$

45. Factor each trinomial in the previous scenario to confirm that it is a perfect square trinomial.

46. Fill in the blanks to make each pair of expressions equal.

a. $\left(x-\frac{5}{2}\right)^2 = x^2 - \_\_\_ x + \_\_\_$

★b. $\left(x+\_\_\_\right)^2 = x^2+3x+\_\_\_$

47. Complete each trinomial to make it a perfect square.

a. $x^2+x+\_\_\_$

b. $x^2-5x+\_\_\_$

★c. $x^2-\frac{3}{2}x+\_\_\_$

48. Factor each trinomial in the previous scenario to confirm that it is a perfect square trinomial.

49. You will now look at another quadratic equation and use some very simple mathematical ideas to discover a way to solve any quadratic equation.

Start with the equation $x^2+2x-5=0$.

The trinomial is not factorable, but you can change it to make it factorable.

Focus on the "–5" for a moment. As long as it occupies the third spot in the trinomial, you will not be able to factor the trinomial. Move the –5 to the other side of the equation.

Now your equation looks like this: $x^2+2x=5$

Actually, write the equation like this: $x^2+2x+\_\_\_=5$

50. Do you see where this is going? In several of the previous scenarios, you worked on completing trinomials to make them become perfect squares. What number would you place in the blank above to create a perfect square trinomial?

51. If the blank spot in $x^2+2x+\_\_\_=5$ is filled with "+1," the left side of the equation is a perfect square trinomial, but you cannot change an equation by simply writing in a number. Can you place a number in an equation without making that equation have a different solution than the original one?

There is a simple way to place a "+1" in the equation. When you add 1 to the left side of the equation, add 1 to the right side also. This will keep the equation balanced and make it look like $x^2+2x+1=5+1$ or, more simply, $x^2+2x+1=6$. Since the left side is now a perfect square trinomial, it can be factored as $(x+1)^2=6$. This is an equation that you already know how to solve.

52. Solve the equation $(x+1)^2=6$.

In the previous scenario, you started with the equation $x^2+2x-5=0$ and used some clever mathematics to create a perfect square trinomial, x^2+2x+1. Then you factored this to create a squared binomial, $(x+1)^2$. From there, you were able to take the square root of both sides and eventually find the original equation's solution: $x=-1\pm\sqrt{6}$. This is the exact solution.

53. Find the approximate values of $-1\pm\sqrt{6}$ by using a calculator.

54. Now try to solve the equation $x^2-6x-1=0$. Start by moving the "-1" to the other side of the equation. You will then have an incomplete trinomial. Refer to the previous scenarios if you need more guidance.

55. Use a calculator to find the approximate values of the previous solution. Round to the nearest tenth.

When you use this new method to solve a quadratic equation, you take a trinomial that cannot be factored and move the third term (the constant term) to make the trinomial incomplete, in a sense. Then, you carefully choose a number to add to both sides of the equation, a number that makes the incomplete trinomial a perfect square. As a result, this method is known as **Completing the Square**.

56. Solve each equation using the method of Completing the Square.

a. $x^2 - 10x + 21 = 0$

b. $x^2 + 8x - 3 = 0$

57. Complete each trinomial to make it a perfect square.

a. $x^2 - 24x + \_\_\_$

b. $x^2 + 7x + \_\_\_$

★c. $x^2 - \frac{2}{3}x + \_\_\_$

58. Factor each trinomial in the previous scenario to confirm that it is a perfect square trinomial.

59. Solve the equation $x^2 - 5x + 3 = 0$ by using the method of Completing the Square.

60. Consider the equation $x^2 - 5x + 3 = 0$ again. If you multiply both sides of the equation by 2, it becomes $2x^2 - 10x + 6 = 0$. It looks different but it has the same solution, just as $x + 2 = 3$ and $2x + 4 = 6$ have the same solutions. Use the method of Completing the Square to solve this equation and try to figure out why it is hard to solve it.

61. Do not solve the following equations. Instead, describe the operation that you need to perform to make it possible to solve the equation using the Completing the Square method.

a. $3x^2 - 24x + \_\_\_ = -3$

b. $-x^2 + 3x + \_\_\_ = -5$

62. Solve the following equations using the Completing the Square method.

a. $3x^2 - 24x + 3 = 0$

b. $-2x^2 + 6x - 12 = -8$

63. Write out a set of directions that you could follow to solve any quadratic equation with the Completing the Square method.

64. How can you determine the x-intercepts of a function without looking at its graph? Use what you describe to determine the x-intercept of the function $y = 2x + 8$.

65. Notice the graph of the function shown to the right. The shape of this function is known as a parabola (pronounced puh-RA-buh-luh). If you tossed a small rock into the parabola, notice where it would settle: at the bottom. The location where this rock would settle is called the vertex of the parabola.

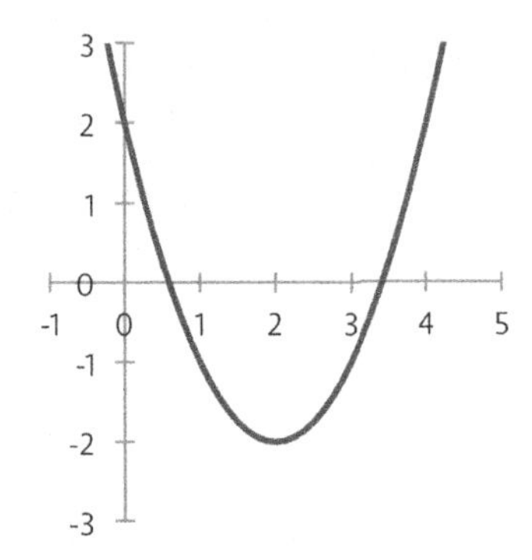

a. Estimate the coordinates of the vertex of this parabola.

b. How many x-intercepts does this function have?

c. Identify the y-intercept of this parabola.

66. Determine the x-intercepts of the function $f(x)=x^2-4x+2$. The graph of this function is shown in the previous scenario, which should help you see if your x-intercepts are accurate.

67. Solve the following equations using the Completing the Square method.

a. $4x^2-32x-44=0$

b. $-x^2+3x+1=0$

68. As you learn other methods, you may forget that you can sometimes solve quadratic equations by factoring. To review that strategy, solve the equation $x^2 - 6x + 8 = 0$ by factoring.

69. Part of a parabola is shown. If the parabola is extended, you will see a second x-intercept. Estimate the location of this second x-intercept.

a.

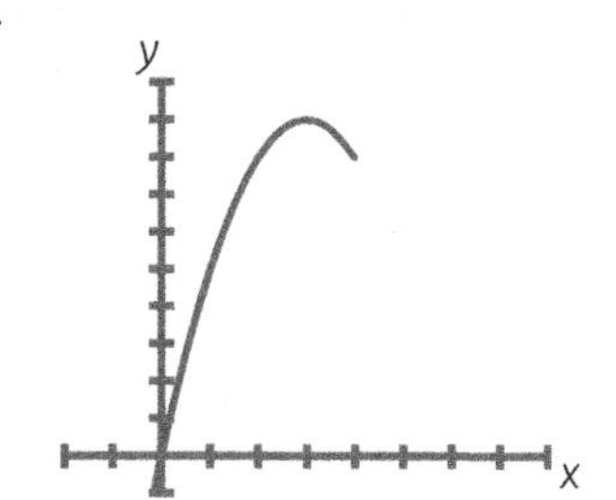

b.

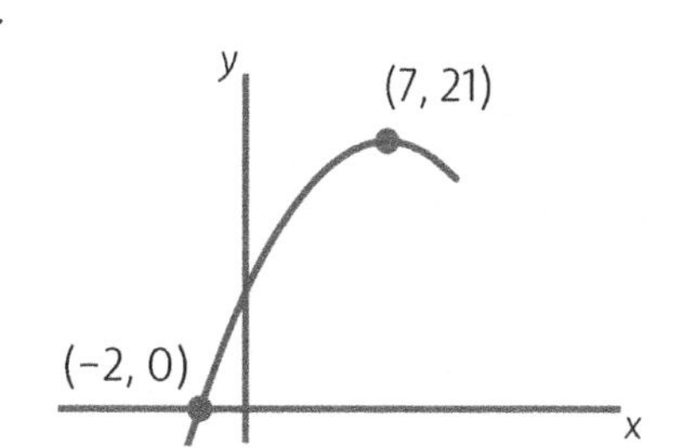

Use this page to record important ideas in the previous section or for any other writing that helps you learn the topics in this book.

Section 7

SOLVING QUADRATIC EQUATIONS WITH THE QUADRATIC FORMULA

Any equation with a quadratic structure can be written as $ax^2+bx+c=0$. You have seen that you can always solve this type of equation by using the method of Completing the Square.

70. ★Solve the equation $ax^2+bx+c=0$ by using the method of Completing the Square. Do this by following the guiding steps below. As you read, actively follow the steps by rewriting them in the space to the right.

a. Start by writing down the initial equation.

$$ax^2+bx+c=0$$

b. Make the leading coefficient 1 by dividing everything by "a."

$$\frac{ax^2}{a}+\frac{bx}{a}+\frac{c}{a}=\frac{0}{a} \rightarrow \text{This can be simplified to become } x^2+\frac{b}{a}x+\frac{c}{a}=0.$$

c. Move the third term of the trinomial to the other side.

$$x^2+\frac{b}{a}x\_\_\_\_=-\frac{c}{a}$$

d. Complete the square, and remember to add to both sides.

$$x^2+\frac{b}{a}x+\left(\frac{b}{2a}\right)^2=-\frac{c}{a}+\left(\frac{b}{2a}\right)^2$$

e. Factor the left side and combine the right side into a single fraction.

$$\left(x+\frac{b}{2a}\right)^2=\frac{b^2-4ac}{4a^2}$$

f. Take the square root of both sides.

$$x+\frac{b}{2a}=\pm\sqrt{\frac{b^2-4ac}{4a^2}} \rightarrow \text{This can be simplified as } x+\frac{b}{2a}=\pm\frac{\sqrt{b^2-4ac}}{2a}.$$

g. Isolate x.

$$x=-\frac{b}{2a}\pm\frac{\sqrt{b^2-4ac}}{2a} \rightarrow \text{As a single fraction, the solution is } x=\_\_\_\_\_\_\_\_\_\_\_\_.$$

71. This fraction is very confusing right now, but it will make more sense later. The idea here is that if an equation has the form $ax^2+bx+c=0$, then the solution will always be $x=$ ____________. Rather than going through the process of Completing the Square, you can take the coefficients of $ax^2+bx+c=0$ (the values of a, b, and c) and plug them into the formula $x=$ ____________. Since this formula allows you to solve any quadratic equation, it is known as The Quadratic Formula.

72. The Quadratic Formula seems complicated at first, but it will become more familiar as you use it to solve quadratic equations. For now, write out The Quadratic Formula.

73. For example, in the equation $2x^2-6x-8=0$, the values of a, b and c are 2, -6 and -8, in that order.

To use the Quadratic Formula to solve the equation, you would write $x=\dfrac{-(-6)\pm\sqrt{(-6)^2-4(2)(-8)}}{2(2)}$.

Carefully simplify this expression to find the solution to the original equation.

a. To start with, try to simplify the expression $(-6)^2-4(2)(-8)$.

b. Now, write the simplified form of the expression $\sqrt{(-6)^2-4(2)(-8)}$.

74. What is the simplified form of the expression $x=\dfrac{-(-6)\pm\sqrt{(-6)^2-4(2)(-8)}}{2(2)}$?

75. Simplify the following expressions.

a. $\dfrac{-(3)\pm\sqrt{(3)^2-4(5)(-2)}}{2(5)}$

b. $\dfrac{-(-2)\pm\sqrt{(-2)^2-4(5)(-1)}}{2(5)}$

76. What is the value of the expression $\dfrac{-b\pm\sqrt{b^2-4ac}}{2a}$ if a = 2, b = 4 and c = 1?

77. What is the value of the expression $\frac{-b\pm\sqrt{b^2-4ac}}{2a}$ if a = 1, b = 3 and c = 4?

78. If you wanted to solve each of the following equations using the Quadratic Formula, what numbers would you use for a, b, and c?

a. $2x^2-3x+1=0$

b. $x^2-8=0$

c. $-5x^2+4x-9=0$

79. If you wanted to solve each of the following equations using the Quadratic Formula, what numbers would you use for a, b, and c?

a. $x^2=-2x+5$

★b. $2x^2-x+3=5x+3$

80. Consider the equation $x^2-2x-3=0$. Use the Quadratic Formula to find the solution to this equation. Simplify the result as much as you can.

81. Now use the Quadratic Formula to find the solution to $-x^2+2x+3=0$.

82. Earlier, you found that the solution to the equation $x^2+2x-5=0$ is $x=-1\pm\sqrt{6}$.

a. Use the Quadratic Formula to solve the equation and confirm this solution.

b. Use a calculator to round $-1\pm\sqrt{6}$ to the nearest tenth.

c. In the equation $x^2+2x-5=0$, if you move the three terms to the other side of the equation, it becomes $0=-x^2-2x+5$, but the solution must still be the same. Use the Quadratic Formula to solve the equation $0=-x^2-2x+5$.

83. Use the Quadratic Formula to find the solution to the equation $x^2+2x+5=0$. Notice the subtle difference between this equation and the equation in the previous scenario.

84. Consider the equation $3x^2-2x-4=0$.

a. Use the Quadratic Formula to confirm that the solution is $x=\dfrac{1\pm\sqrt{13}}{3}$.

b. Use a calculator to determine the value of $\dfrac{1\pm\sqrt{13}}{3}$ rounded to the nearest tenth.

85. Determine the x-intercepts of the parabola represented by the function $f(x) = -\frac{1}{2}x^2 + x + 3$.

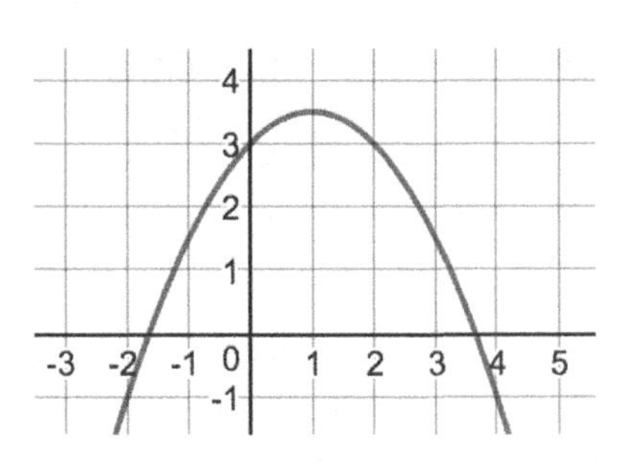

a. Determine the exact value of the x -intercepts.

b. Use a calculator to approximate the value of the x-intercepts to the nearest tenth. Refer to the graph to verify that your results are accurate.

c. Estimate the coordinates of the vertex of this parabola.

86. Locate the exact x-intercepts of the function $g(x) = -\frac{1}{2}x^2 + x - 3$.

Use this page to record important ideas in the previous section or for any other writing that helps you learn the topics in this book.

Section 8
REVIEW

87. Solve each equation.

a. $x^2 = 20$

b. $x^2 = -\frac{1}{100}$

88. Solve each equation by factoring.

a. $x^2 - 25 = 0$

b. $x^2 - 3x = 0$

89. Find the value of the expressions below, rounded to the nearest tenth.

a. $\frac{-2 \pm \sqrt{13}}{3}$

b. $x = \frac{1}{7} \pm \frac{2\sqrt{5}}{7}$

90. Complete the trinomial to make it a perfect square trinomial. Check your result by factoring.

a. $x^2 + 10x +$ ____

b. $x^2 + 5x +$ ____

★c. $x^2 - \frac{3}{2}x +$ ____

91. Without changing it, can this equation be solved using the Quadratic Formula? If it can, what values would you use for a, b, and c? Do not solve the equation.

$-2x^2 + 5x - 7 = 6$

92. Without changing it, can this equation be solved using the Quadratic Formula? If it can, what values would you use for a, b, and c? Do not solve the equation.

$\frac{1}{3}x^2 + 2x - 1 = 0$

93. Use the Quadratic Formula to solve the equation $-\frac{1}{3}x^2 + 2x - 1 = 0$.

Use this page to record important ideas in the previous section or
for any other writing that helps you learn the topics in this book.

Section 9

THE VERTEX OF A PARABOLA

In two earlier scenarios, you found the x–intercepts of a function, and a graph of the function was provided to help confirm that your results were accurate. The following scenarios will look closely at the graphs of parabolas to help you become more familiar with their unique shape.

94. Draw a graph for each function below, using a T-chart. Include at least 6 points in your graph.

a. $y=x^2$

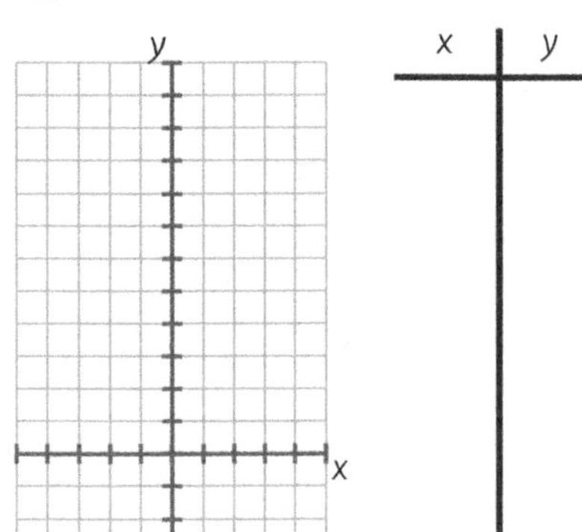

b. $y=-x^2$

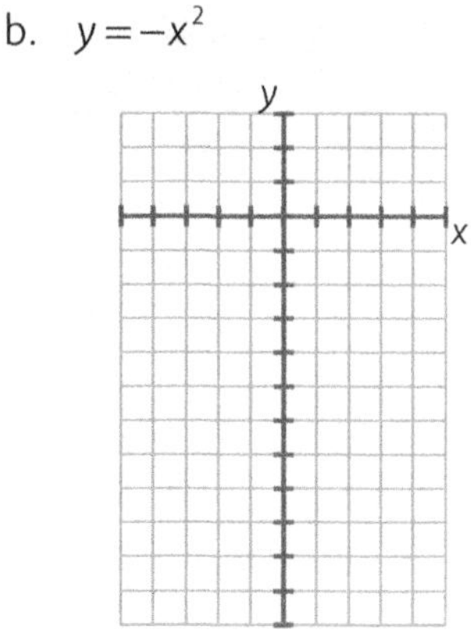

95. Identify the coordinates of the vertex of each parabola in the previous scenario.

96. Look at each of the three parabolas below. Write the coordinates of the x-intercepts and then write the coordinates of the vertex.

a.

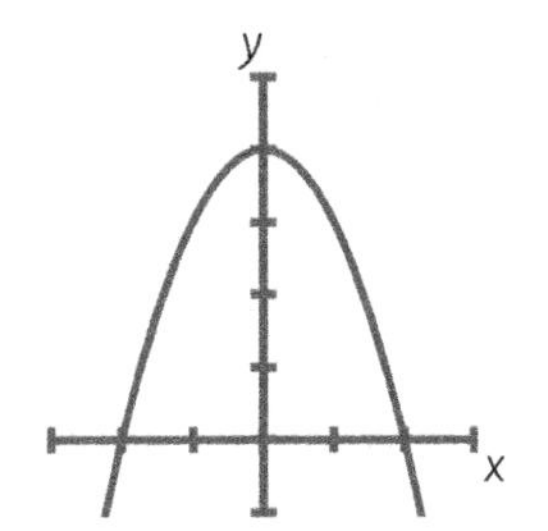

b.

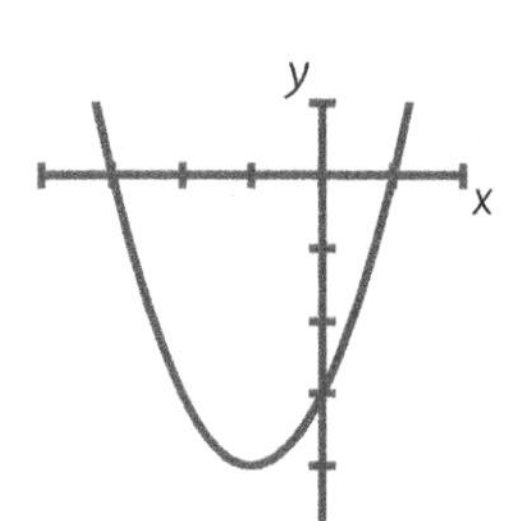

c.

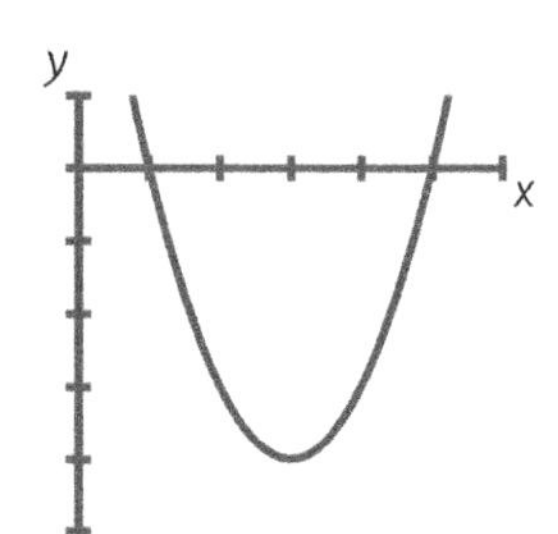

97. In the previous scenario, compare the x-value of the vertex and the location of the x-intercepts.

98. If the x-intercepts of a parabola are (4, 0) and (6, 0), what is the x-value of the vertex?

99. What is the x-value of the vertex in the parabola shown in the graph?

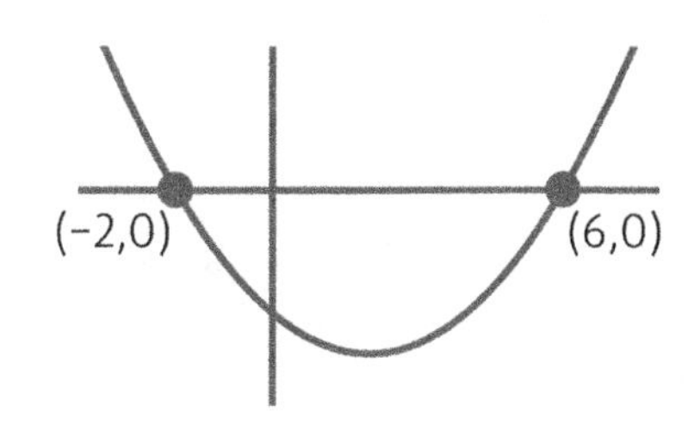

100. What is the x-value of the vertex of a parabola if its x-intercepts are located at the points shown?

a. (0, 0) and (6, 0) b. (−5, 0) and (3, 0) c. (4, 0) and (−1, 0)

101. What is the x-value of the vertex of a parabola if its x-intercepts are located at the points shown?

a. (A, 0) and (B, 0) ★b. $\left(2+\sqrt{3},\ 0\right)$ and $\left(2-\sqrt{3},\ 0\right)$

102. If the vertex of a parabola is located at (4, 9) and one of the x-intercepts is located at (1, 0), where is the other x-intercept located?

One feature that makes a parabola special is its symmetry. Each parabola can be cut in half along its line of symmetry, which passes through the vertex. As a result, knowing the location of the vertex helps determine other points along the parabola.

103. One way to find the vertex is to use a parabola's symmetry. First, locate the x-intercepts. Then, find the x-value halfway between them. For the function below, determine the x-value of the vertex.

$y = x^2 - 4x - 5$

104. For each function below, determine the x-value of the vertex.

a. $f(x) = x^2 - 16$ ★b. $g(x) = x^2 - 3x - 10$

105. What number is halfway between. . .

a. 4 + 1 and 4 - 1? b. 4 + 2 and 4 - 2? c. 4 ± 3? d. $4 \pm \sqrt{3}$?

106. Write the number that is halfway between the two numbers shown.

a. $\sqrt{2}$ and $-\sqrt{2}$ b. $1+\sqrt{2}$ and $1-\sqrt{2}$ c. $\frac{1}{3}+\frac{\sqrt{2}}{3}$ and $\frac{1}{3}-\frac{\sqrt{2}}{3}$

107. Write the number that is halfway between the two numbers shown.

a. $3 \pm \sqrt{5}$ b. $A \pm \sqrt{B}$ c. $-\frac{a}{2} \pm \frac{1}{2}$ d. $\frac{F}{G} \pm \frac{\sqrt{H}}{G}$

108. For a quadratic function in the form $y = ax^2 + bx + c$, you can locate the x-intercepts by solving the equation $0 = ax^2 + bx + c$. If you use the Quadratic Formula to solve this equation, the solution is $x = \frac{-b \pm \sqrt{b^2 - 4ac}}{2a}$. Separate this fraction to make it $\frac{-b}{2a} \pm \frac{\sqrt{b^2 - 4ac}}{2a}$.

a. What number is halfway between $\frac{-b}{2a} \pm \frac{\sqrt{b^2 - 4ac}}{2a}$?

b. Since the vertex of a parabola is always located halfway between the x-intercepts, the vertex will always have an x-value of _______.

109. If a = 1 and b = 6, what is the value of $\frac{-b}{2a}$?

110. What is the value of $\frac{-b}{2a}$ if a and b are assigned the values shown below?

a. a = 1 and b = -8

b. a = 2 and b = 3

★c. $a = \frac{1}{3}$ and b = -4

111. For a quadratic function in the form $y=ax^2+bx+c$, you can locate the x-intercepts by solving the equation $0=ax^2+bx+c$. If you use the Quadratic Formula to solve this equation, the solution is $x=\frac{-b\pm\sqrt{b^2-4ac}}{2a}$. Separate this fraction to make it $\frac{-b}{2a} \pm \frac{\sqrt{b^2-4ac}}{2a}$. Since the vertex of a parabola is always located halfway between the x-intercepts, the vertex will always have an x-value of ______.

112. For each function below, determine the x-value of the vertex by substituting the function's coefficients of a and b into the ratio $x=\frac{-b}{2a}$.

a. $y=x^2+6x+2$

b. $y=2x^2-8x-7$

c. $f(x)=-x^2+16$

113. ★What is the x-value of the vertex of the parabola formed by the function $g(x)=\frac{1}{2}x^2-\frac{3}{2}x-5$?

114. If you only know the x-value of the vertex, you only have half of the location of that point. You still need the y-value before you can plot the vertex point.

a. If you have the equation for a parabola, how can you find the y-value of its vertex, if you only know the x-value?

b. For the equation $y=x^2-4x+1$, the vertex is located at (2, ___). Fill in the blank.

115. Find the coordinates of the vertex for each quadratic function shown below. Use $x=\frac{-b}{2a}$ to find the x-value. Then use the function to find the y-value.

a. $y=x^2+4x-2$

b. $y=-2x^2-8x+1$

c. $y=x^2-9$

116. ★Locate the coordinates of the vertex for the quadratic function shown.

$$f(x)=7+3x+\frac{1}{4}x^2$$

117. ★The vertex of the function $y=x^2-8x+3$ is located at $\left(\frac{-b}{2a}, Y\right)$. What is the value of Y?

118. Part of a parabola has already been graphed for you. Use what you have learned about the symmetry of a parabola to complete each graph.

a.

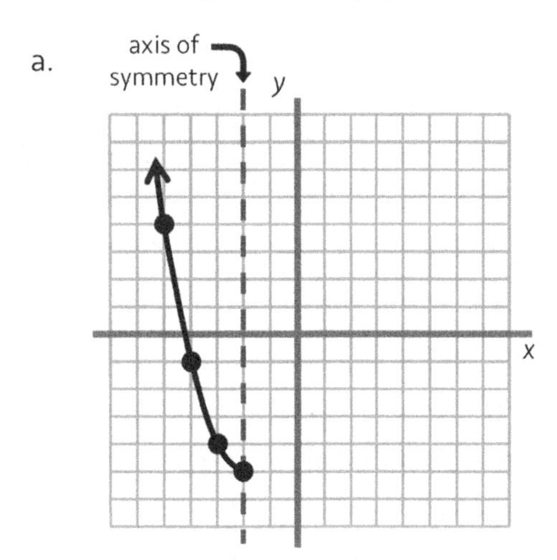

b.

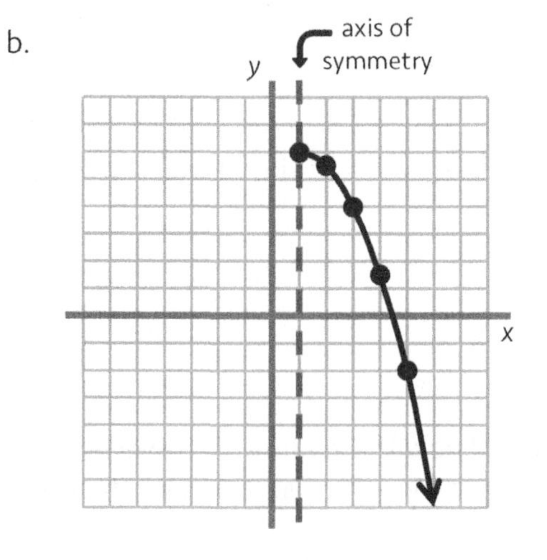

119. In the previous graphs, the axis of symmetry is marked for you. This line is drawn to help you use the symmetry of the parabola to find more points. It is a dashed line to show that it is not actually a visible line. Draw the axis of symmetry on the graph of each parabola below.

a.

b.

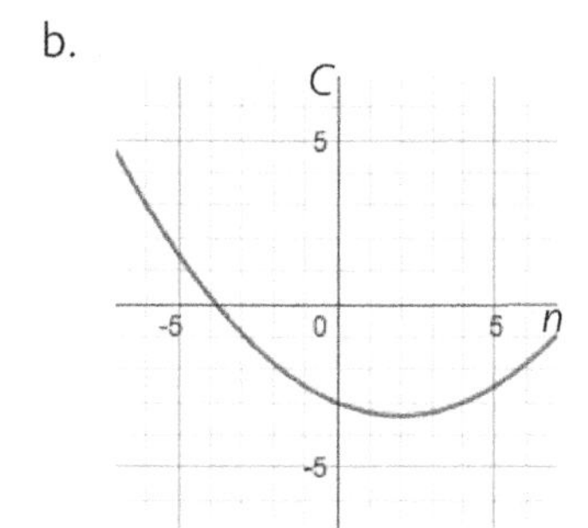

c.

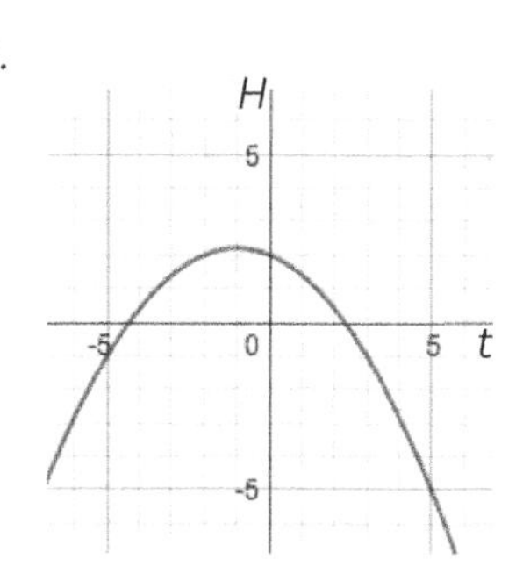

120. Write the equation for the line of symmetry that you drew in each graph in the previous scenario.

121. Estimate the coordinates of the vertex for each parabola in the previous scenario.

122. How can you use the vertex to find a parabola's axis of symmetry?

123. What is the equation for the axis of symmetry of the graph of $y=x^2+10x-13$?

124. To prepare for the next section, simplify each expression below.

a. $(-3)^2$ b. $-3(-2)^2$ c. $-(-1)^2$ d. $2(-2)^2-3(2)-5$

Use this page to record important ideas in the previous section or for any other writing that helps you learn the topics in this book.

Section 10

GRAPHING A PARABOLA

To graph a parabola, you can plug in x-values until you find the general shape of the curve. A possibly better method, though, is to start by finding the vertex of the parabola. It is useful to know the location of the vertex because the vertex splits the parabola into its two symmetric curves.

125. Graph the function $y = x^2 - 2x - 3$ by finding points as shown below.

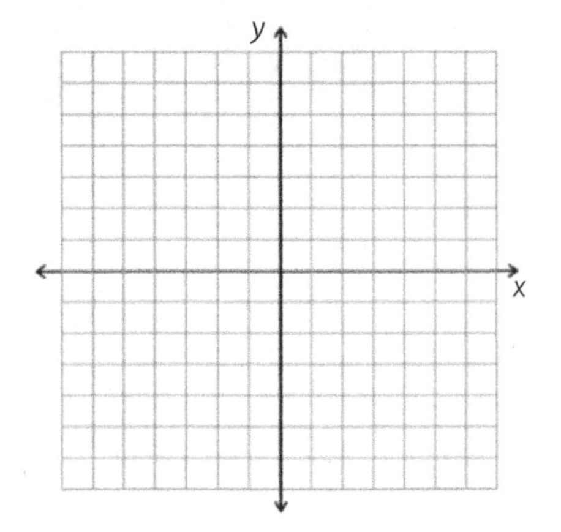

a. Find the vertex.

b. Find the y-intercept by letting $x = 0$.

c. Find the x-intercepts by letting $y = 0$.

d. Plot other points by plugging in other x-values. Pick x-values close to the points you have already found.

e. Draw a smooth parabola through the points that you have plotted on the graph.

126. Graph the function $y=-x^2-2x+3$ by finding points in the order below.

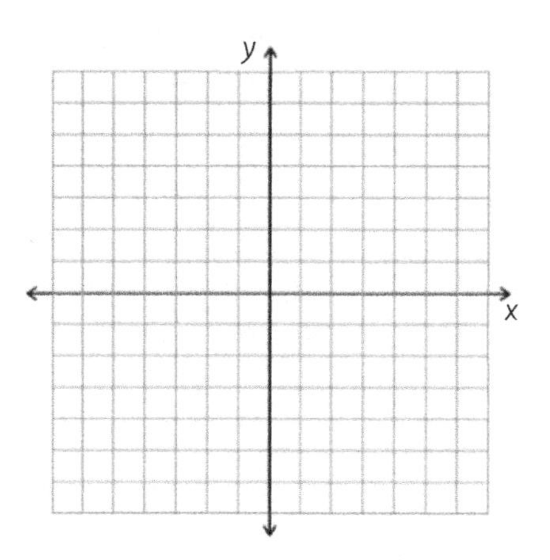

a. Find the vertex.

b. Find the y-intercept by letting $x = 0$.

c. Find the x-intercepts by letting $y = 0$.

d. Plot other points by plugging in other x-values.

e. Draw a smooth parabola through the points.

127. Graph the parabola by using the strategy listed in the previous scenario. Plot at least six points.

$y=x^2-4$

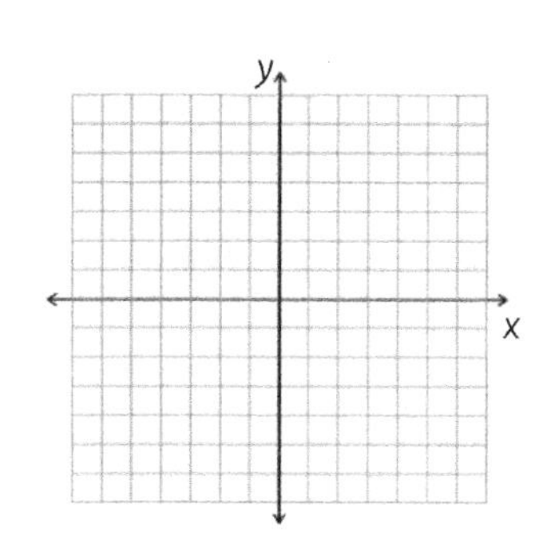

128. Graph the parabola by using the strategy listed in the previous scenario. Plot at least six points.

$$y = -\frac{1}{2}x^2 + 2x + 3$$

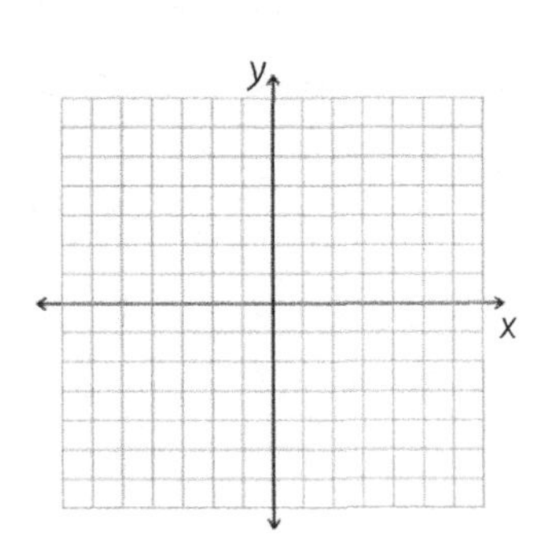

129. Draw a basic sketch of a parabola that matches the description below.

a. The vertex is also the y-intercept.

b. The vertex is also the x-intercept.

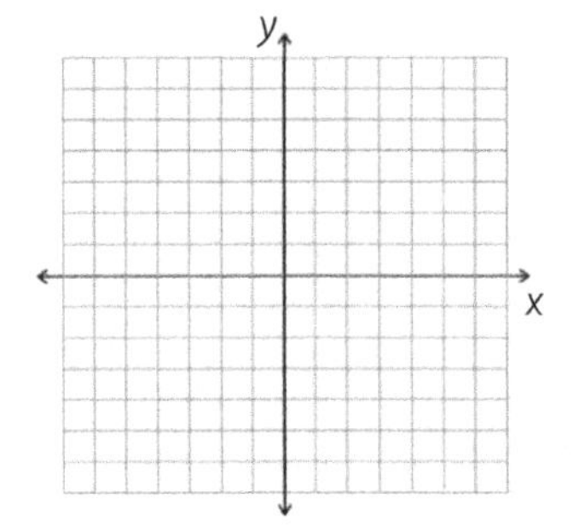

c. The vertex is also the x- and y-intercept.

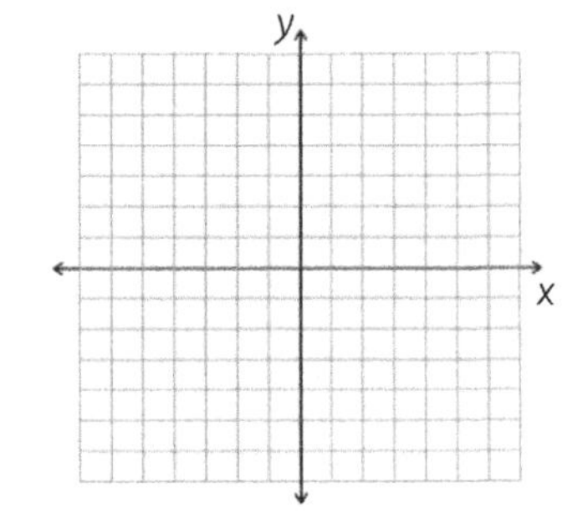

130. ★Graph the parabola by plotting at least seven points.

$$f(x) = -\frac{1}{4}x^2 + x - 1$$

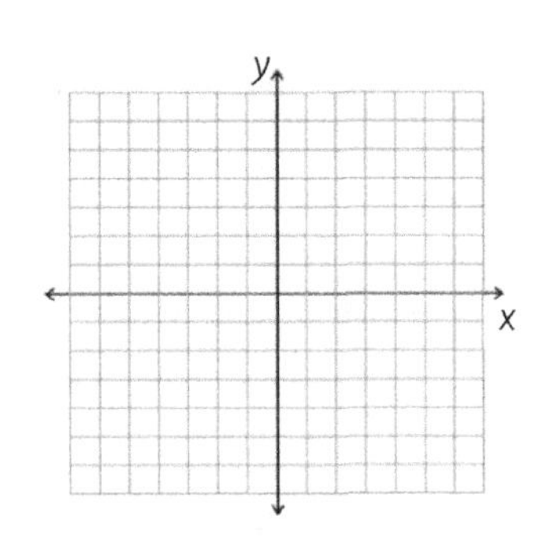

131. Identify the vertex and x-intercepts of the parabola modeled by $f(x)=5-2x+x^2$.

132. Make the equation match the graph shown by filling in the blank.

$y=-x^2+\_\_\_x-5$

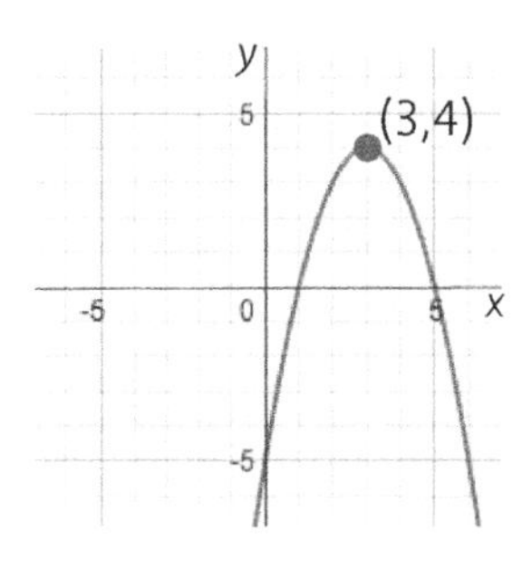

133. ★Make the equation match the graph below it by filling in the blank.

a. $C=\frac{1}{10}n^2+\_\_\_n-3$

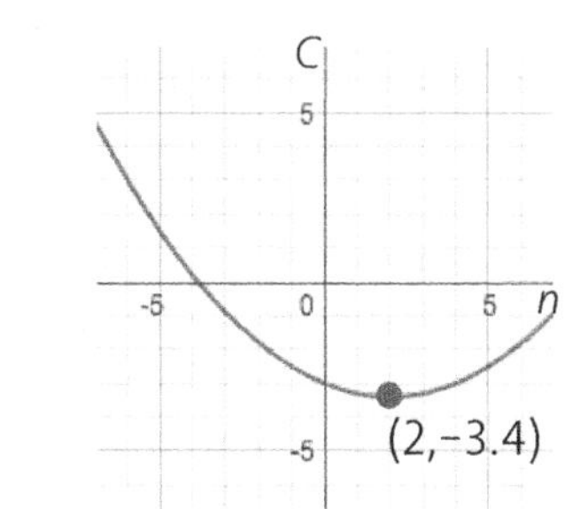

b. $H=-\frac{1}{5}t^2+\_\_\_t+2$

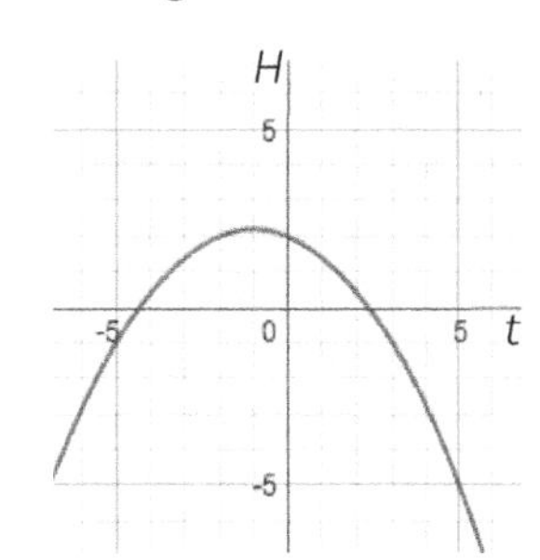

When you graph a quadratic function, the shape of the parabola will roughly resemble either ∪ or ∩. If its shape is ∪, it opens upward because, like a cup, the opening is up. If its shape is ∩, it opens downward. Imagine tipping a cup upside down when it is full of water. The water would fall downward to the floor. If you haven't realized it yet, there is a simple way to determine whether a parabola will open upward or downward. Go back through all of the scenarios that contain a parabola and its equation and see if you can figure out this relationship.

134. How can you use a parabola's equation to determine if it opens upward or downward?

135. Look at each function below and state whether the parabola will open upward or downward.

a. $f(x)=-3x^2+x-7$ b. $g(n)=\frac{1}{4}n^2-4n$ c. $H(t)=3-0.05t^2$

136. Draw a very basic sketch of a parabola that has each of the following characteristics.

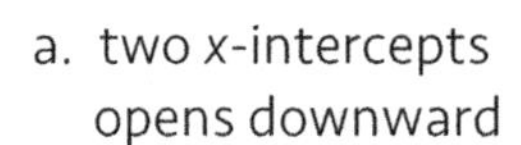
a. two x-intercepts opens downward

b. one x-intercept opens upward

c. no x-intercepts opens downward

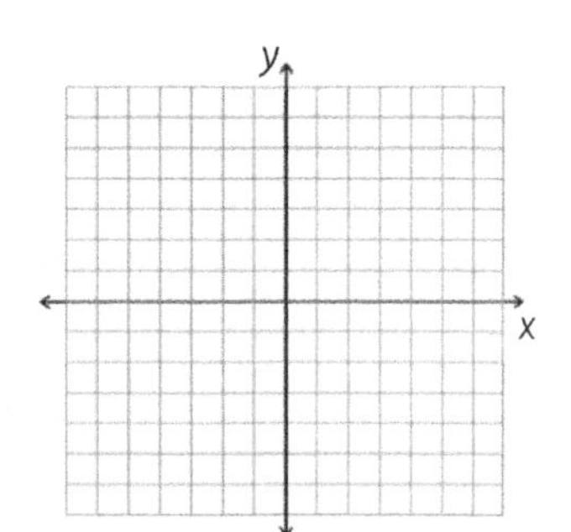

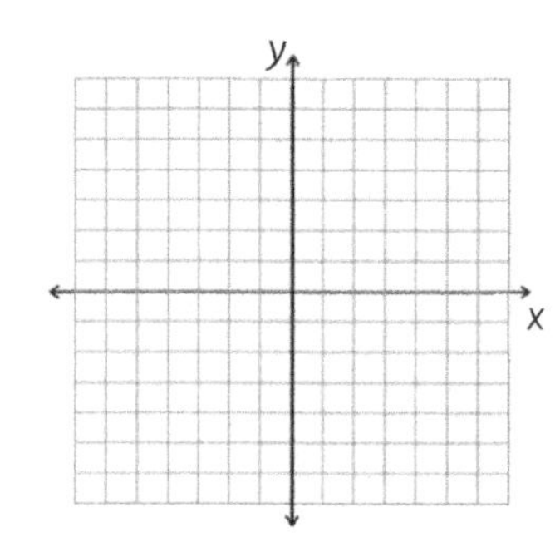

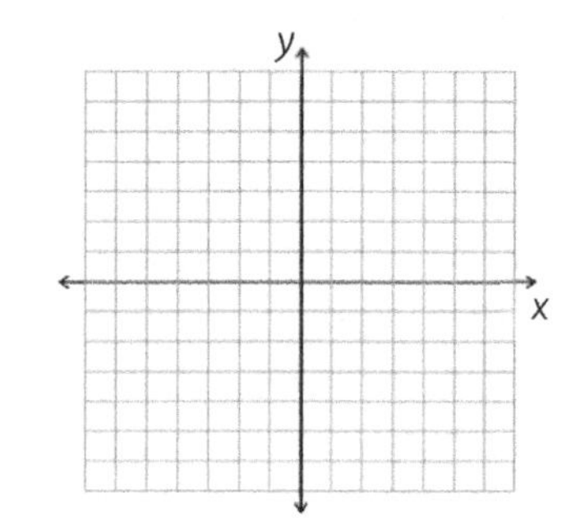

137. Simplify each expression.

a. $\dfrac{-4}{2\left(\frac{1}{5}\right)}$ b. $\dfrac{-(-5)}{2\left(\frac{1}{8}\right)}$

138. If $f(n)=-\frac{1}{2}n^2-n+3$, what is the value of each expression below?

a. $f(0)$ b. $f(2)$ c. $f(-1)$

Use this page to record important ideas in the previous section or for any other writing that helps you learn the topics in this book.

Section 11

SCENARIOS THAT INVOLVE QUADRATIC FUNCTIONS

139. A baseball is thrown through the air from a position 6 feet above the ground. The height H in feet after time t in seconds is given by the formula $H(t) = -2t^2 + 4t + 6$.

a. What is the maximum height of the baseball?

b. How long does it take to reach the maximum height?

c. When does the baseball hit the ground?

140. A stone is launched from a platform above the ground. The height of the stone t seconds after its launch is given by $H(t) = -5t^2 + 20t + 10$, where H is the height, measured in meters.

a. After how many seconds does the stone reach the maximum height?

b. What is the maximum height of the stone?

c. After how many seconds does the stone land on the ground?

d. How high above the ground is the stone at the moment it is launched?

141. Consider the equation $H=16t^2+3t+4$, where H is the height of a ball, in feet, after it has been in the air for t seconds. Could this equation model the path that a ball follows when it is tossed from one person to another person?

142. A bug is crawling along a branch and slips on a patch of slime. It falls to the ground below and its path is modeled by the function $H=-16t^2+16$, where H is the height above the ground, in feet, and t is how much time the bug has been falling through the air, in seconds.

a. For how many seconds did the bug fall before it landed on the ground?

b. What was the height of the bug at the moment it started to fall?

143. The eruption of a volcano sends ash and rocks flying through the air. The graph shows the path of one of these rocks from the moment it leaves the top of the volcano until it hits the ground.

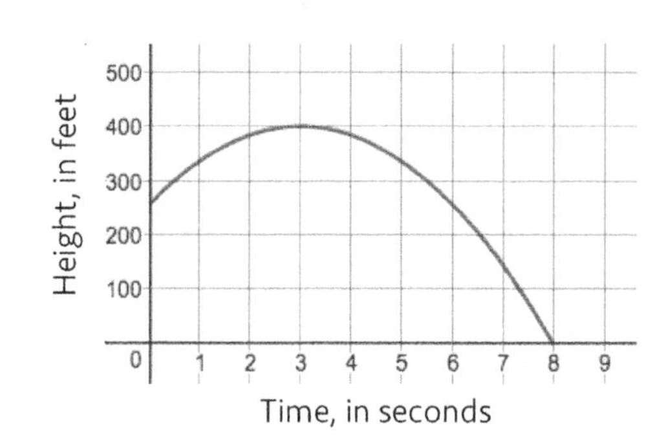

a. How many seconds is the rock in the air?

b. The rock is at its highest point after ____ seconds.

c. Does the rock reach its maximum height halfway through its flight through the air? Explain why this happens.

144. In the previous scenario, for how many seconds is the rock above 300 feet as it travels through the air? Estimate without doing numerical calculations.

145. ★A deer stands in the middle of the road and does not move. To avoid hitting the deer, a driver presses the brakes to make the car slow down quickly. As the car slows down, the speed of the car is modeled by the formula $S(t) = -0.6t^2 - 9t + 60$, where S is the speed of the car, in miles per hour, after the driver has been pressing the brakes for a total of t seconds.

   a. How fast was the car moving at the moment the driver started pressing the brakes?

   b. How many seconds does the car travel before the car stops?

Use this page to record important ideas in the previous section or for any other writing that helps you learn the topics in this book.

Section 12

GRAPHING QUADRATIC INEQUALITIES

146. Bring your brain back to a previous topic: graphing linear inequalities. Graph the following inequalities in the Cartesian Plane provided.

a. $y \geq \frac{2}{3}x - 1$

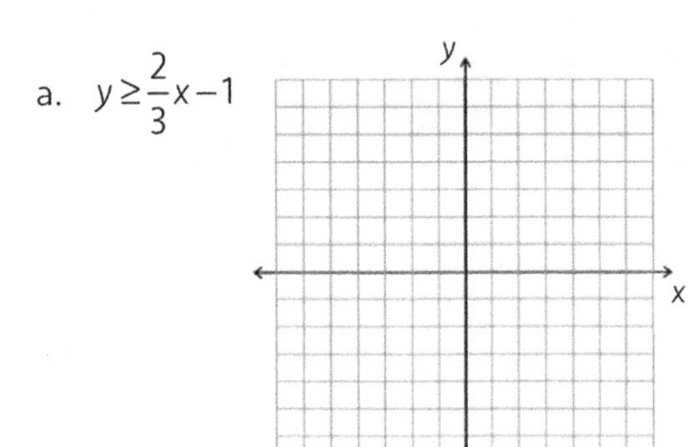

b. $y < 3x + 2$

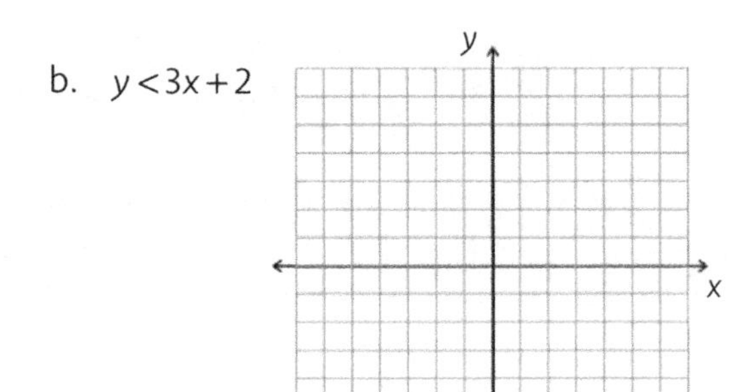

147. Graph the following quadratic inequalities. You have never done this before, but you can figure out how to do this by using what you know about graphing the linear inequalities in the previous scenario.

a. $y < x^2 - 4$

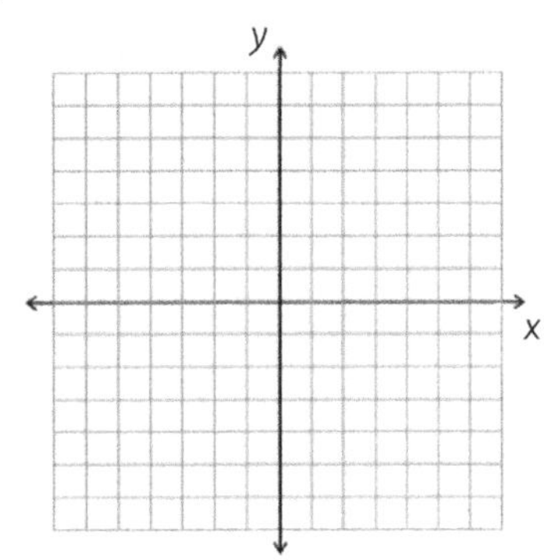

b. $y \geq x^2 - 3x - 4$

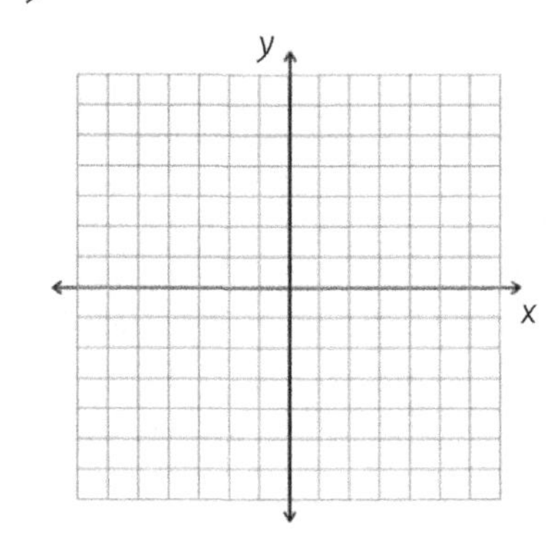

148. ★Graph the following quadratic inequality.

$y \geq -\frac{1}{2}x^2 - x + 4$

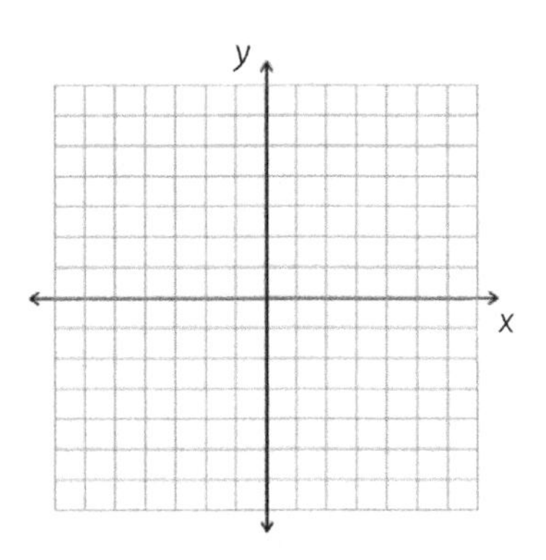

Use this page to record important ideas in the previous section or for any other writing that helps you learn the topics in this book.

Section 13
CUMULATIVE REVIEW

149. Use the graph to find the value of each expression listed below. Estimate if necessary.

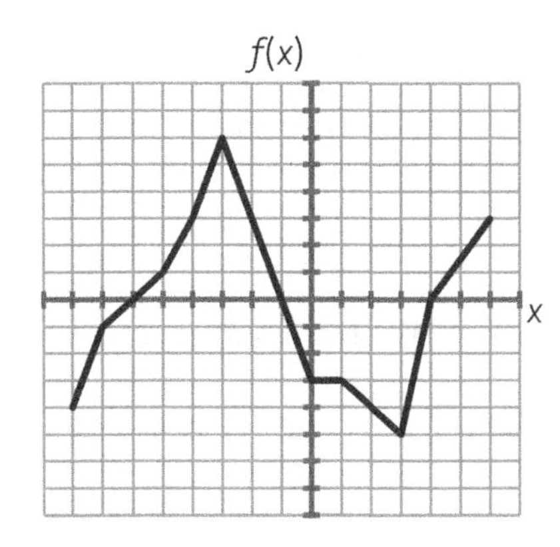

a. $f(6)$

b. x when $f(x) = 0$

150. What are the domain and range of the function shown in the previous scenario?

151. Given: $f(x)=x+3$ and $g(x)=\sqrt{x}$.

a. Evaluate $g(-9)$.

b. What is $g[f(22)]$?

152. The basketball court in the gym is a rectangle, 84 feet long and 50 feet wide. To the nearest foot, what is the distance between opposite corners of the court?

153. If a supporting beam for a bridge is constructed using squares, as shown in the image below, it will warp and lose its strength over time. If the opposite corners of each square are connected with a diagonal beam, the bridge will be stronger, and it will last longer. In the diagram below, draw one diagonal beam in each square. How long is each diagonal beam that connects opposite corners? Write the length in simplest radical form and then round the length to the nearest tenth of a meter.

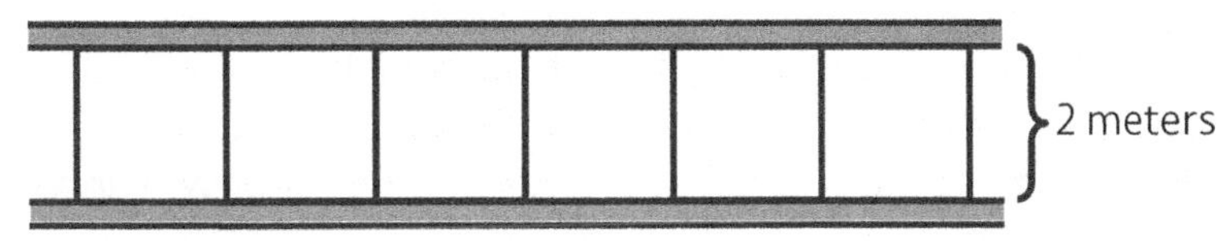

154. If $\sqrt{A}=3\sqrt{5}$, what is the value of A? If $\sqrt{B}=\frac{3\sqrt{2}}{2}$, what is the value of B?

155. If $\sqrt{H}=\sqrt{2}+\sqrt{2}+\sqrt{2}+\sqrt{2}$, determine the value of H.

156. ★Determine the value of $A + B$ if $\sqrt{A}+\sqrt{B}=\sqrt{48}+\sqrt{72}-\sqrt{12}-2\sqrt{8}$.

157. Determine the equation of the line that passes through the points $\left(\sqrt{8}, \sqrt{6}\right)$ and $\left(\sqrt{18}, \sqrt{24}\right)$. Write your equation in Slope-Intercept Form.

158. Use what you know about simplifying radicals to verify that the expression below is equal to 1.

$$\frac{\sqrt{3}\cdot\sqrt{30}}{\sqrt{5}\left(\sqrt{50}-\sqrt{8}\right)}$$

159. While flying a kite at a park with the sun directly overhead, you determine that you are 30 feet from the kite's shadow. You notice that you have let out the maximum amount of string in your spool, which is 90 feet. How high above the ground is the kite? Write your answer in a.) simplest radical form and b.) rounded to the nearest foot.

Use this page to record important ideas in the previous section or
for any other writing that helps you learn the topics in this book.

Section 14
ANSWER KEY

1.	No. Let x = 350 and solve for $H \rightarrow H$ = -172. Since the height is negative, the ball has already landed on the ground after it has traveled 350 feet.
2.	Solve the equation $0=-0.01x^2+3x+3$. In upcoming scenarios, you will learn how to solve this type of equation (factoring is not possible).
3.	a. 11 b. 3; 13x c. 7, 5
4.	a. 4 b. 5 c. -10
5.	a. $(x-6)(x-2)$ b. $(x-6)(x+6)$ c. $(x-7)(x-7)$
6.	a. $(2x+3)(x+2)$ b. $(3x+7)(2x-5)$ c. $(3x+2)(3x+2)$
7.	a. $x=-1,-9$ b. $x=7,-7$ c. $x=\frac{5}{2}$
8.	Make the y-value(s) of the function equal 0 and solve the resulting equation to find all values of x that have a y-value of 0.
9.	(-5,0) and (2,0) Solve $0=x^2+3x-10$.
10.	a. 3 b. 7 c. 11
11.	a. $3\sqrt{3}$ b. $2\sqrt{6}$ c. $4\sqrt{5}$
12.	a. $\frac{2}{3}$ b. $\frac{\sqrt{5}}{3}$ c. $\frac{2\sqrt{2}}{3}$ d. 1 e. $\sqrt{3}$
13.	$\frac{3}{\sqrt{5}}\cdot\frac{\sqrt{5}}{\sqrt{5}}=\frac{3\sqrt{5}}{5}$
14.	a. $\frac{2}{\sqrt{3}}\cdot\frac{\sqrt{3}}{\sqrt{3}}=\frac{2\sqrt{3}}{3}$ b. $\frac{10}{\sqrt{2}}\cdot\frac{\sqrt{2}}{\sqrt{2}}=\frac{10\sqrt{2}}{2}=5\sqrt{2}$ b. $\frac{7\sqrt{3}}{\sqrt{7}}\cdot\frac{\sqrt{7}}{\sqrt{7}}=\frac{7\sqrt{21}}{7}=\sqrt{21}$
15.	a. $\frac{4}{5}$ b. $\frac{\sqrt{3}}{2}$ c. $\frac{2}{\sqrt{3}}\rightarrow\frac{2}{\sqrt{3}}\cdot\frac{\sqrt{3}}{\sqrt{3}}\rightarrow\frac{2\sqrt{3}}{3}$
16.	It is not possible...? Or is it? Hmmm....
17.	–
18.	i
19.	$2i$
20.	$3i$

21.	a. 4 b. 7 c. -81 d. $\frac{2}{3}$ e. $-\frac{1}{4}$
22.	It makes it clear that the i is not <u>under</u> the square root symbol.
23.	a. $2i\sqrt{2}$ b. $2i\sqrt{5}$ c. $2i\sqrt{11}$
24.	a. $\frac{4i}{5}$ b. $\frac{i\sqrt{3}}{2}$ c. $\frac{i}{\sqrt{2}}$ or $\frac{i\sqrt{2}}{2}$
25.	$i^2=-1$ (keep reading for an explanation)
26.	a. $3^2=9$ b. $7^2=49$ c. $\left(2\sqrt{6}\right)^2=24$ d. $\sqrt{\frac{4}{9}}=\frac{2}{3}\rightarrow\left(\frac{2}{3}\right)^2=\frac{4}{9}$ e. $\sqrt{-1}=i\rightarrow i^2=-1$
27.	a. $4i^2\rightarrow4(-1)\rightarrow-4$ b. $4i^2\rightarrow4(-1)\rightarrow-4$ c. $25i^2\rightarrow25(-1)\rightarrow-25$ d. -3
28.	a. $x=5$ or $-5\rightarrow x=\pm5$ The symbol $\pm$ reads as "plus or minus" and is used to write "5 or -5" in a shorter form. b. $x^2=9\rightarrow x=\pm3$ c. $x=\pm4$
29.	Squaring a positive number gives the same result as squaring a negative number, so there are 2 possible solutions.
30.	a. $x+1=\pm3$ b. $x-7=\pm5$
31.	$\sqrt{(x-4)^2}=\sqrt{3}\rightarrow x-4=\pm\sqrt{3}\rightarrow x=4\pm\sqrt{3}$
32.	$4+\sqrt{3}\approx5.7$ $4-\sqrt{3}\approx2.3$
33.	$\sqrt{(x-1)^2}=\sqrt{25}\rightarrow x-1=\pm5\rightarrow x=6$ or -4
34.	a. $x=-11\pm9\rightarrow x=-20$ or -2 b. $x=7\pm\sqrt{2}$
35.	divide both sides by 2
36.	$\sqrt{(x+4)^2}=\sqrt{9}\rightarrow x+4=\pm3\rightarrow x=-1$ or -7
37.	a. $x=-11\pm\sqrt{5}$ b. $x=5\pm\sqrt{2}$
38.	a. $x=\pm3i$ b. $x=\pm i\sqrt{5}$
39.	a. $(3i)^2=9i^2=9\cdot-1=-9$; $(-3i)^2=9i^2=-9$

	b. $\left(i\sqrt{5}\right)^2=5i^2=5\cdot-1=-5$; $\left(-i\sqrt{5}\right)^2=5i^2=-5$
40.	a. 2 b. 4; 4
41.	a. 9 b. 4; 16 c. 7; -14
42.	a. 1; 1 b. 2; 4 c. 8; 64
43.	a. $B=2A$ b. $C=A^2$ c. $C=\left(\frac{B}{2}\right)^2$
44.	a. 36 b. 1 c. 64
45.	a. $(x+6)^2$ b. $(x-1)^2$ c. $(x-8)^2$
46.	a. 5; $\frac{25}{4}$ b. $\frac{3}{2}$; $\frac{9}{4}$
47.	a. $\frac{1}{4}$ b. $\frac{25}{4}$ c. $\frac{9}{16}$
48.	a. $\left(x+\frac{1}{2}\right)^2$ b. $\left(x-\frac{5}{2}\right)^2$ c. $\left(x-\frac{3}{4}\right)^2$
49.	–
50.	+1
51.	If you add 1 to both sides
52.	$x=-1\pm\sqrt{6}$
53.	1.4, -3.4
54.	$x^2-6x\_\_\_\_=1\to x^2-6x\underline{+9}=1\underline{+9}$ $\to(x-3)^2=10\to x-3=\pm\sqrt{10}\to x=3\pm\sqrt{10}$
55.	6.2, -0.2
56.	a. $x=5\pm2\to3$ or 7 b. $x=-4\pm\sqrt{19}$
57.	a. 144 b. $\frac{49}{4}$ c. $\frac{1}{9}$
58.	a. $(x-12)^2$ b. $\left(x+\frac{7}{2}\right)^2$ c. $\left(x-\frac{1}{3}\right)^2$
59.	$x=\frac{5}{2}\pm\frac{\sqrt{13}}{2}$
60.	$2x^2$ cannot be the first term of a perfect square trinomial, at least, not the ones you are comfortable with for now.
61.	a. Divide both sides by 3 b. Divide both sides by -1
62.	a. $x=4\pm\sqrt{15}$ b. $x=\frac{3}{2}\pm\frac{1}{2}\to x=2$ or 1
63.	1. Make the leading coefficient 1. 2. Move the third term of the trinomial to the other side. 3. Compute $\left(\frac{b}{2}\right)^2$ to complete the square, and remember to add to both sides. 4. Factor the left side and combine the right side into a single number. 5. Take the square root of both sides. Remember the "plus or minus" $\to\pm$. 6. Isolate x.

64.	Let y = 0 and solve for x. The x-int. is (-4,0).
65.	a. (2,-2) b. Two c. (0,2)
66.	Let y = 0 and solve for x. $x=2\pm\sqrt{2}\to3.4$ or $0.6\to(3.4, 0)$ and $(0.6, 0)$
67.	a. $4\pm3\sqrt{3}$ b. $x=\frac{3}{2}\pm\frac{\sqrt{13}}{2}$
68.	$(x-4)(x-2)=0\to x=4$ or 2
69.	a. (6, 0) b. (16, 0)
70.	g. $\frac{-b\pm\sqrt{b^2-4ac}}{2a}$
71.	In both of the blanks shown, you should write the expression $\frac{-b\pm\sqrt{b^2-4ac}}{2a}$.
72.	$x=\frac{-b\pm\sqrt{b^2-4ac}}{2a}$
73.	a. $36+64\to100$ b. $\sqrt{100}\to10$
74.	$x=\frac{6\pm10}{4}\to x=\frac{16}{4}$ or $\frac{-4}{4}\to x=4$ or -1
75.	a. $\frac{-3\pm7}{10}\to-1, \frac{2}{5}$ b. $\frac{2\pm\sqrt{24}}{10}\to\frac{2\pm2\sqrt{6}}{10}\to\frac{1\pm\sqrt{6}}{5}$
76.	$\frac{-(4)\pm\sqrt{(4)^2-4(2)(1)}}{2(2)}\to\frac{-4\pm\sqrt{8}}{4}\to$ $\frac{-4\pm2\sqrt{2}}{4}\to\frac{-2\pm\sqrt{2}}{2}$ or $-1\pm\frac{\sqrt{2}}{2}$
77.	$\frac{-3\pm\sqrt{(3)^2-4(1)(4)}}{2(1)}\to\frac{-3\pm\sqrt{-7}}{2}\to\frac{-3\pm i\sqrt{7}}{2}$
78.	a. a = 2, b = -3, c = 1 b. a = 1, b = 0, c = -8 c. a = -5, b = 4, c = -9
79.	a. a = 1, b = 2, c = -5 b. a = 2, b = -6, c = 0
80.	$\frac{2\pm4}{2}$, which simplifies to become -1 or 3
81.	$\frac{-2\pm4}{-2}$, which is also equivalent to -1 or 3
82.	a. – b. -3.4, 1.4 c. same solution:

	$\frac{1\pm\sqrt{6}}{-1}$ is equal to $\frac{-1\pm\sqrt{6}}{1}$
83.	$x=-1\pm 2i$
84.	a. $\frac{2\pm\sqrt{4+48}}{6}\to\frac{2\pm2\sqrt{13}}{6}\to\frac{1\pm\sqrt{13}}{3}$ b. x = 1.5, −0.9
85.	a. $\left(1+\sqrt{7},0\right)$ and $\left(1-\sqrt{7},0\right)$ b. (3.6, 0) and (−1.6, 0) c. (1, 3.5)
86.	$\left(1+i\sqrt{5},0\right)$ and $\left(1-i\sqrt{5},0\right)$ These results contain imaginary numbers so they do not exist on a real number line or in the Cartesian Plane. Thus, $g(x)$ does not cross the x-axis (no x-intercepts).
87.	a. $x=\pm2\sqrt{5}$ b. $x=\pm\frac{1}{10}i$
88.	a. $x=5,\ -5$ b. $x=0,\ 3$
89.	You need a calculator to do this. a. x = 0.5, −1.9 b. x = 0.8, −0.5
90.	a. $x^2+10x+25$ $(x+5)^2$ b. $x^2+5x+\frac{25}{4}$ $\left(x+\frac{5}{2}\right)^2$ c. $x^2-\frac{3}{2}x+\frac{9}{16}$ $\left(x-\frac{3}{4}\right)^2$
91.	No, the equation must equal "0".
92.	Yes, a = $\frac{1}{3}$, b = 2, and c = −1. However, it would be easier to use the Quadratic Formula if you first cleared the fraction in $\frac{1}{3}x^2$ by multiplying both sides of the equation by 3.
93.	First, multiply both sides of the equation by −3 to make it $x^2-6x+3=0$. $\frac{6\pm\sqrt{(-6)^2-4(1)(3)}}{2(1)}\to\frac{6\pm\sqrt{24}}{2}$ $\to\frac{6\pm2\sqrt{6}}{2}\to3\pm\sqrt{6}$

94.	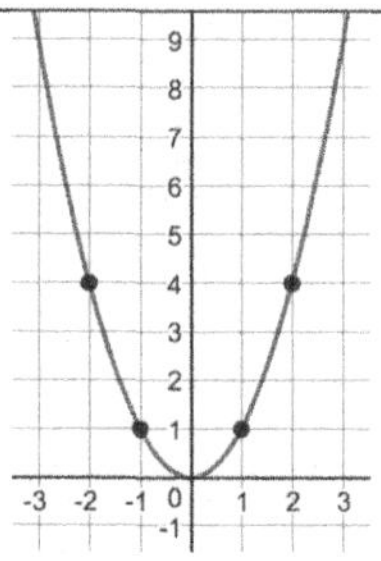 a. 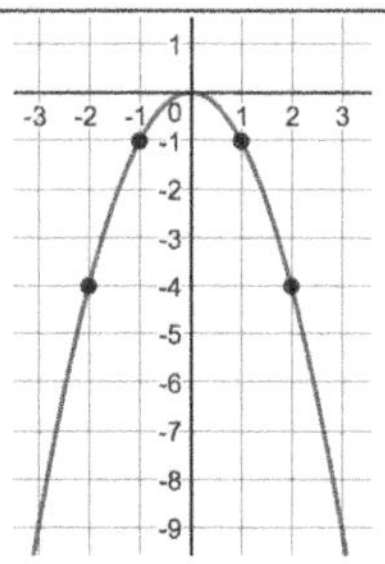 b.
95.	both vertices are located at (0, 0)
96.	a. x-int: (−2,0), (2,0); vertex: (0,4) b. x-int: (−3,0), (1,0); vertex: (−1,−4) c. x-int = (1,0), (5,0); vertex: (3,−4)
97.	The x-value of the vertex is halfway between the x-intercepts
98.	5
99.	x = 2 (2 is halfway between −2 and 6)
100.	a. 3 b. −1 c. 1.5
101.	a. $\frac{A+B}{2}$ b. 2
102.	(7, 0)
103.	Solve $0=x^2-4x-5\to0=(x-5)(x+1)$ $\to x=5,\ -1\to$ 2 is halfway between 5 and −1 The x-value of the vertex is 2.
104.	a. 0; the x-ints are (4,0) and (−4,0) b. 1.5; the x-ints are (5,0) and (−2,0)
105.	a-d. 4
106.	a. 0 b. 1 c. $1/3$
107.	a. 3 b. A c. $-\frac{a}{2}$ d. $\frac{F}{G}$
108.	a. $\frac{-b}{2a}$ b. $\frac{-b}{2a}$
109.	$\frac{-6}{2(1)}=-3$
110.	a. $\frac{-(-8)}{2(1)}=4$ b. $\frac{-(3)}{2(2)}=-\frac{3}{4}$ c. $\frac{-(-4)}{2\left(\frac{1}{3}\right)}=4\div\frac{2}{3}=4\cdot\frac{3}{2}=6$
111.	$\frac{-b}{2a}$
112.	a. $x=\frac{-6}{2(1)}=-3$ b. $x=\frac{-(-8)}{2(2)}=2$ c. $x=\frac{-(0)}{2(-1)}=0$

113.	$x=\dfrac{-\left(-\frac{3}{2}\right)}{2\left(\frac{1}{2}\right)}=\dfrac{3}{2}$
114.	a. plug x-value into the function b. −3
115.	a. $x=-2$; $y=-6$ b. $x=-2$; $y=9$ c. $x=0$; $y=-9$
116.	$x=-6$; $y=-2$
117.	−13
118.	a. b.
119.	a. b. c.
120.	a. $x=3$ b. $x=2$ c. $x=-1$
121.	a. (3, 4) b. (2, −3-ish) c. (−1, 2-ish)
122.	The x-value of the vertex is the same as the axis of symmetry.
123.	$x=-5$; find the vertex: $x=\dfrac{-10}{2(1)}=-5$
124.	a. 9 b. $-3\cdot4\rightarrow-12$ c. $-(1)\rightarrow-1$

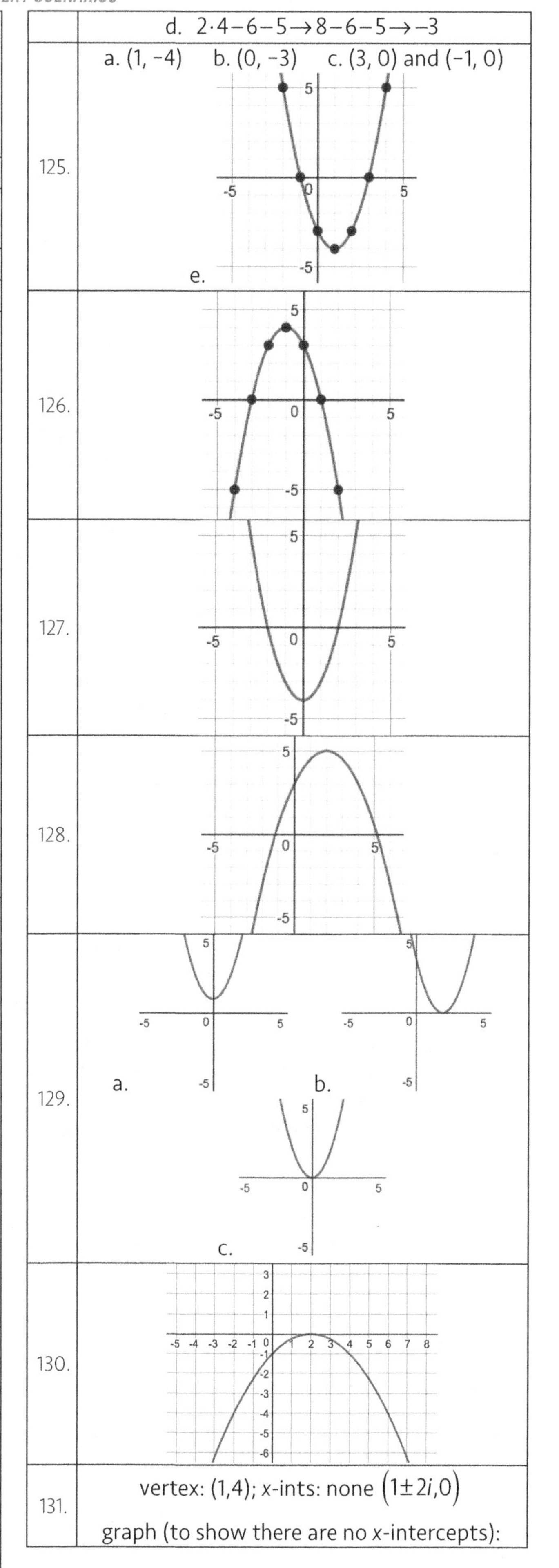

	d. $2\cdot4-6-5\rightarrow8-6-5\rightarrow-3$
125.	a. (1, −4) b. (0, −3) c. (3, 0) and (−1, 0) e.
126.	
127.	
128.	
129.	a. b. c.
130.	
131.	vertex: (1,4); x-ints: none $(1\pm2i,0)$ graph (to show there are no x-intercepts):

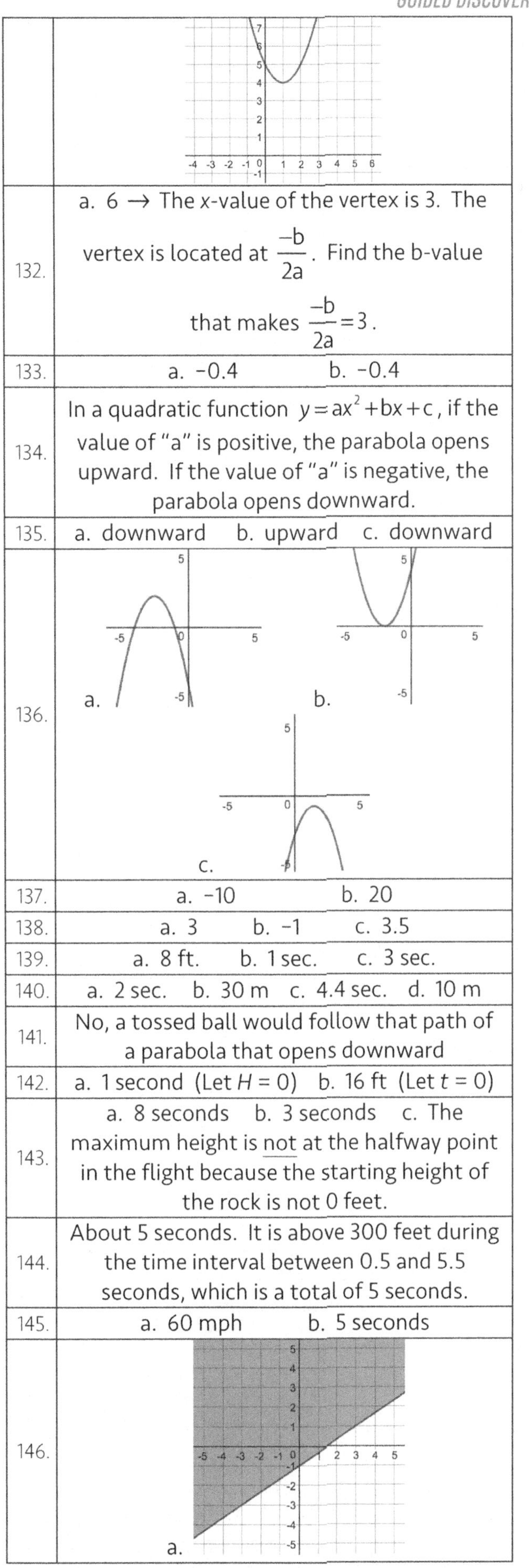

	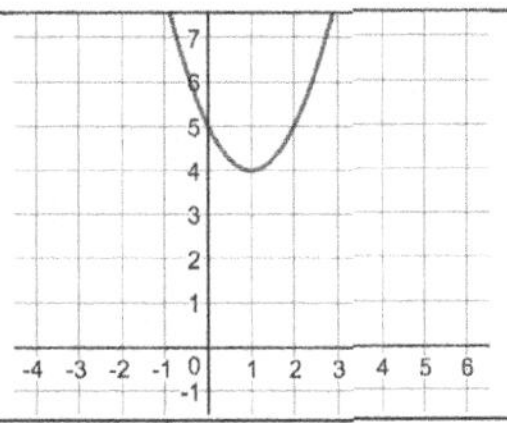
132.	a. 6 → The *x*-value of the vertex is 3. The vertex is located at $\frac{-b}{2a}$. Find the b-value that makes $\frac{-b}{2a}=3$.
133.	a. −0.4 b. −0.4
134.	In a quadratic function $y=ax^2+bx+c$, if the value of "a" is positive, the parabola opens upward. If the value of "a" is negative, the parabola opens downward.
135.	a. downward b. upward c. downward
136.	a. 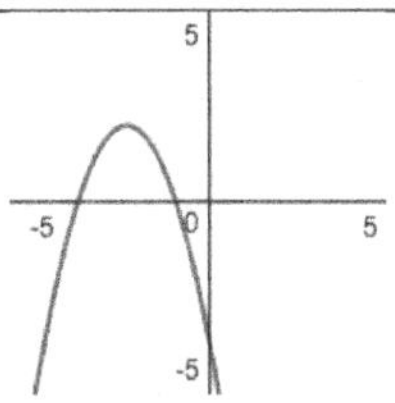 b. 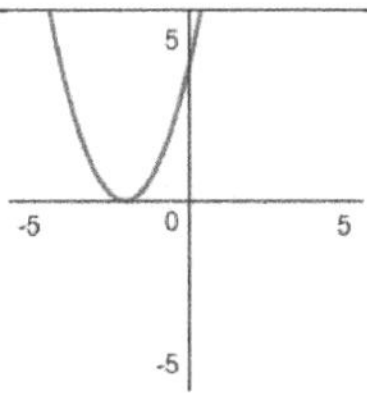 c. 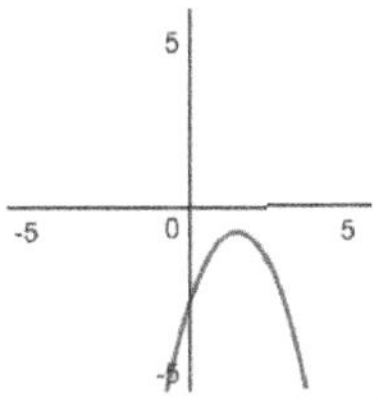
137.	a. −10 b. 20
138.	a. 3 b. −1 c. 3.5
139.	a. 8 ft. b. 1 sec. c. 3 sec.
140.	a. 2 sec. b. 30 m c. 4.4 sec. d. 10 m
141.	No, a tossed ball would follow that path of a parabola that opens downward
142.	a. 1 second (Let *H* = 0) b. 16 ft (Let *t* = 0)
143.	a. 8 seconds b. 3 seconds c. The maximum height is not at the halfway point in the flight because the starting height of the rock is not 0 feet.
144.	About 5 seconds. It is above 300 feet during the time interval between 0.5 and 5.5 seconds, which is a total of 5 seconds.
145.	a. 60 mph b. 5 seconds
146.	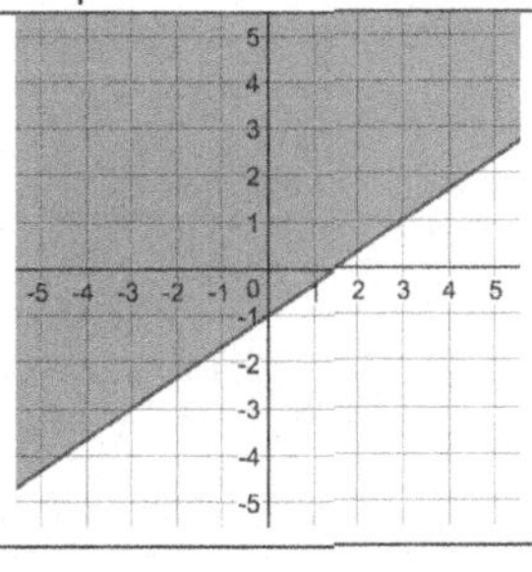 a.

	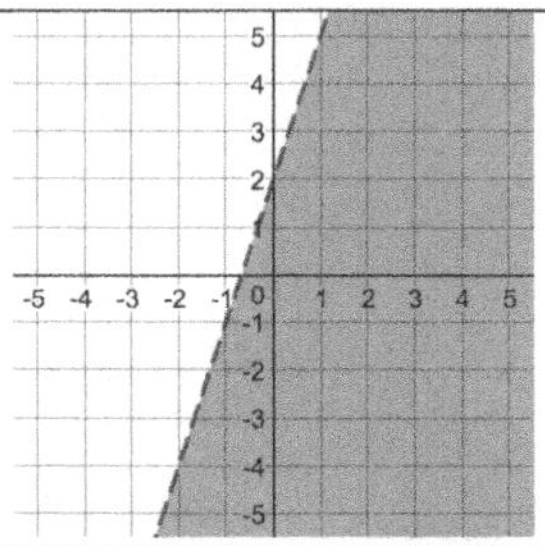 b.
147.	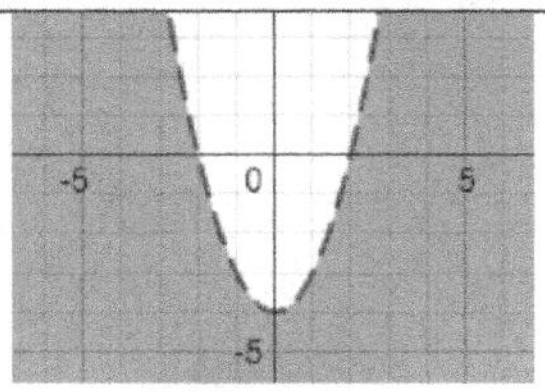 a. 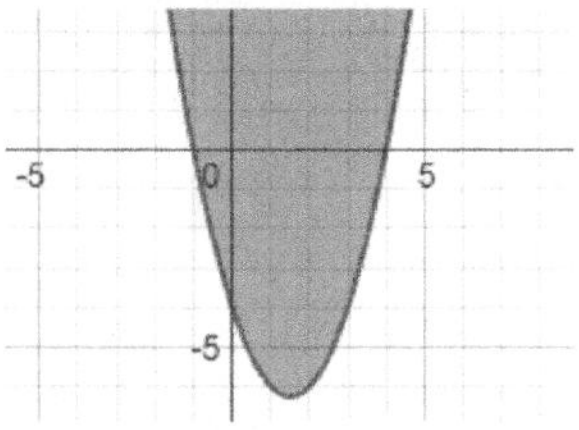 b.
148.	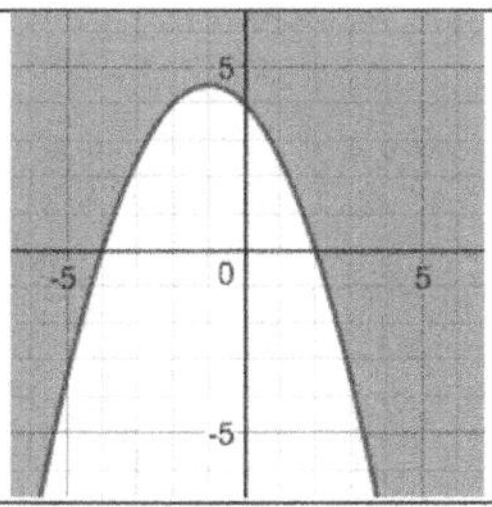
149.	a. 3 b. $x=-6,-1,4$
150.	a. domain: $-8\le x\le 6$ b. range: $-5\le f(x)\le 6$
151.	a. $g(-9)=\sqrt{-9}=3i$ b. $f(22)=22+3=25\rightarrow g(25)=\sqrt{25}=5$
152.	98 feet
153.	$2\sqrt{2}$ or 2.8 m
154.	$A=45$ $B=\frac{9}{2}$ or 4.5
155.	$H=32$
156.	$A+B=20$
157.	$y=\sqrt{3}x-\sqrt{6}$
158.	$\frac{\sqrt{90}}{\sqrt{5}\left(5\sqrt{2}-2\sqrt{2}\right)}\rightarrow\frac{3\sqrt{10}}{\sqrt{5}\left(3\sqrt{2}\right)}\rightarrow\frac{3\sqrt{10}}{3\sqrt{10}}\rightarrow 1$
159.	solve: $30^2+b^2=90^2\rightarrow b^2=7200\rightarrow b=\sqrt{7200}\rightarrow 60\sqrt{2}$ or ≈ 85 feet

HOMEWORK & EXTRA PRACTICE SCENARIOS

As you complete scenarios in this part of the book, you will practice what you learned in the guided discovery sections. You will develop a greater proficiency with the vocabulary, symbols and concepts presented in this book. Practice will improve your ability to retain these ideas and skills over longer periods of time.

There is an Answer Key at the end of this part of the book. Check the Answer Key after every scenario to ensure that you are accurately practicing what you have learned. If you struggle to complete any scenarios, try to find someone who can guide you through them.

CONTENTS

Section 1
REVIEW

1. Simplify the following expressions as much as you can.

a. $\frac{3}{4} \div 2$

b. $2 \div \frac{3}{4}$

c. $\frac{3}{4} \div \frac{1}{2}$

★d. $\frac{\frac{1}{2}}{\frac{3}{4}}$

2. Simplify each expression.

a. $(-9)^2$

b. $-2(2)^2$

★c. $1-(-7)^2$

3. A function is shown in the graph.

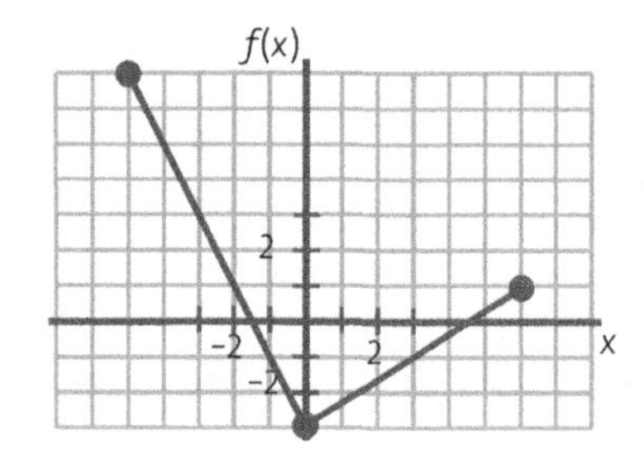

a. What is $f(0)$?

b. Fill in the blank. $f(\_\_\_)=-1$

c. Which value is greater, $f(-4)$ or $f(-3)$?

4. Refer to the graph in the previous scenario.

a. Identify the interval on which the function is increasing. Write the interval as a compound inequality.

b. On what interval is the function decreasing?

5. What is the value of y if $x=-\frac{1}{2}$ for equation shown below?

a. $y=-x^2$

b. $2y-3x=5$

6. Write out the mathematical function below, using the same letters and words that you would say if you were reading the function to another person.

$$f(x)=3x^2+5x-1$$

7. Use the function in the previous scenario to find the value of each expression below.

a. $f(0)$ b. $f(2)$ c. $f(-2)$

8. Change each fraction below to make its denominator a rational number.

a. $\frac{1}{\sqrt{7}}$ b. $\frac{\sqrt{7}}{\sqrt{2}}$ ★c. $\frac{3\sqrt{5}}{\sqrt{2}}$

9. At a high school track and field meet, Ilana sends a shot put flying through the air. As of April 12, 2014, the girls national high school shot put record was 56.7 feet. After it is measured, Ilana's throw does not break the record. Later, Ilana claims that she broke the record and insists that she has video footage to prove it. In the video, the path of Ilana's shot can be modeled by the function $H=-0.02d^2+d+7$, where H is the height of the shot, in feet, and d is the distance the shot has traveled horizontally, also in feet. What could you do to see if the function supports her claim?

10. In the previous scenario, what is $H(0)$ and what does it mean in the context of the scenario?

Section 2

FACTORING REVIEW

11. Fill in the blank to make each expression equal to the factors written below the expression.

a. $x^2 + \_\_\_\_ x + 5$

$(x-5)(x-1)$

b. $-3x^2 + \_\_\_\_ x + 36$

$-3(x+6)(x-2)$

★c. $2x^3 - 2x^2 - 24x$

$2x(x + \_\_\_\_)(x-4)$

12. Factor each of the following expressions.

a. $x^2 - 11x + 28$

b. $2x^2 + 5x - 12$

★c. $9x^2 - 12x + 4$

13. Solve each equation.

a. $x^2 + 9x + 14 = 0$

b. $x^2 - 25 = 0$

★c. $9x^2 - 18x + 9 = 0$

14. Does the function $f(x) = 8 - 4x$ cross the x-axis? If so, what are the coordinates of this point?

15. Where does the function $f(x) = 2x^2 - 18$ cross the x-axis?

Section 3

RADICAL EXPRESSIONS AND IMAGINARY NUMBERS

As you learn how to solve quadratic equations using methods other than factoring, you will need to be familiar with radical expressions, so it will be helpful to review them now.

16. Simplify each expression.

a. $\sqrt{81}$ b. $\sqrt{16}$ c. $\sqrt{144}$

17. Simply each radical expression. As a reminder, $\sqrt{20}$ can be written as $\sqrt{4}\sqrt{5}$, or $2\sqrt{5}$.

a. $\sqrt{45}$ b. $\sqrt{40}$ c. $\sqrt{50}$

18. Simplify each expression.

a. $\sqrt{\frac{16}{25}}$ b. $\sqrt{\frac{6}{25}}$ c. $\sqrt{\frac{24}{25}}$ d. $\sqrt{\frac{0}{25}}$ ★e. $\sqrt{\frac{50}{25}}$

19. What expression do mathematicians use to represent $\sqrt{-1}$?

20. What word does the letter *i* represent in mathematics?

21. What is the value of $\sqrt{-25}$?

22. Find the value of $\bigcup$ that makes each statement true.

a. $\sqrt{-100} = \bigcup i$ b. $\sqrt{-36} = \bigcup i$ c. $\sqrt{-\frac{9}{25}} = \bigcup i$ d. $\sqrt{\bigcup} = \frac{1}{6}i$

23. Simplify each expression.

a. $\sqrt{-20}$ b. $\sqrt{-40}$ c. $\sqrt{-80}$

24. Simplify the square root of each fraction below.

a. $\sqrt{\frac{9}{36}}$ b. $\sqrt{\frac{11}{16}}$ c. $\sqrt{\frac{2}{3}}$

25. Write each expression in simplified form.

a. $\sqrt{-\frac{9}{36}}$ b. $\sqrt{-\frac{24}{25}}$ c. $\sqrt{-\frac{4}{7}}$

26. Since the value of $\sqrt{-1}$ is i, what is the value of i^2?

27. Simplify each expression below.

a. $(3i)^2$ b. $(-3i)^2$ c. $(-6i)^2$ d. $(-i\sqrt{5})^2$

28. A square has an area of 25 square inches. How long is the perimeter of the square?

29. The floor of a kitchen is a square. The floor has a total area of 200 square feet.

a. How long is one side of the kitchen?

b. What is the perimeter of the floor, rounded to the nearest tenth of a foot?

Section 4

QUADRATIC EQUATIONS

30. Solve the following equations. It will help to remember that an equation is considered to be solved when you have identified all possible values that make the original equation true.

a. $w^2 = 9$ b. $3x^2 = 33$ c. $-3 + 4y^2 = 0$

31. Why do the previous equations each have two solutions?

32. Solve each equation shown, which have similar structures to those in the previous scenario.

a. $(x-4)^2 = 1$ b. $(x-1)^2 = 13$ c. $\left(x+\frac{1}{3}\right)^2 = \frac{2}{9}$

33. Isolate x in the following equation: $(x+11)^2 = R$.

34. One final grouping of equations is shown below. Solve each equation.

a. $4(x-3)^2 = 8$ b. $16(x+7)^2 = -1$ c. $9\left(x-\frac{4}{3}\right)^2 = 16$

35. Use what you have learned in previous scenarios to solve the following equations.

a. $x^2 = -1$ b. $x^2 = -10$ c. $x^2 = -\frac{16}{25}$

Section 5

SOLVING QUADRATIC EQUATIONS BY COMPLETING THE SQUARE

Before you learn a new method for solving quadratic equations, you first need to review the process known as squaring a binomial, which shows that an expression like $(x+5)^2$ or $(x+5)(x+5)$ is equivalent to the expression $x^2+10x+25$. There is a distinct relationship between the numbers in these two expressions and you will analyze this relationship in the next scenario.

36. Fill in the blanks below to make each pair of expression equivalent.

a. $(x+3)^2 = x^2 + \_\_\_ x + 9$

b. $(x-11)^2 = x^2 - \_\_\_ x + \_\_\_$

37. Fill in the blanks below to make each pair of expressions equal.

a. $(x + \_\_\_)^2 = x^2 + 14x + \_\_\_$

b. $(x - \_\_\_)^2 = x^2 - x + \_\_\_$

38. Suppose the expression $(x+A)^2$ is equivalent to the expression x^2+Bx+C.

a. What is the relationship between A and B?

b. What is the relationship between B and C?

39. Fill in each blank with a number that makes each expression a perfect square trinomial.

a. $x^2+10x+\_\_\_$

b. $x^2-4x+\_\_\_$

c. $x^2-20x+\_\_\_$

40. Factor each trinomial in the previous scenario to confirm that it is a perfect square trinomial.

41. Complete each trinomial to make it a perfect square.

a. $x^2+24x+\_\_\_$

b. $x^2+5x+\_\_\_$

c. $x^2-7x+\_\_\_$

42. Factor each trinomial in the previous scenario to confirm that it is a perfect square trinomial.

When you use the method of Completing the Square to solve a quadratic equation, you take a trinomial that cannot be factored and remove the third term (the constant term) to make the trinomial incomplete, in a sense. Then, you carefully choose a number to add to both sides of the equation, a number that makes the incomplete trinomial a perfect square.

43. Suppose you are solving the equation below using the method of Completing the Square. What is the next operation you should do to solve the equation?

$$x^2 - 14x - 2 = 13$$

44. Finish solving the equation in the previous scenario.

45. The equation below can be solved by Completing the Square. Do not solve the equation. Instead, describe the next operation you would perform.

$$\begin{aligned} x^2 + 12x - 5 &= -1 \\ +5 \quad & \; +5 \\ x^2 + 12x \quad &= 4 \end{aligned}$$

46. Finish solving the equation in the previous scenario.

47. The equation below can be solved by factoring. The solutions are $x = 3$ and $x = -5$. Use the method of Completing the Square to solve the equation and confirm that those solutions are accurate.

$$x^2 + 2x - 15 = 0$$

48. Solve the equation below by using the method of Completing the Square.

$$x^2 + 4x + 1 = 0$$

49. Use a calculator to write the previous solution in its decimal form. Round to the nearest hundredth.

50. Use the Completing the Square method to solve the equation below.

$$x^2 - 12x + 10 = 2$$

51. Do not solve the following equation. Instead, describe the operation that you would need to perform to make it possible to solve the equation using the Completing the Square method.

$$-4x^2 + 4x - 8 = 0$$

52. Solve the following equations using the Completing the Square method.

a. $2x^2 - 12x = 6$

b. $-x^2 - 10x - 1 = 6$

53. Use a calculator to write each of the previous solutions in their decimal form.

54. Brief Review.

a. How can you determine the x-intercepts of a function without looking at its graph?

b. Determine the x-intercept of the function $y = \frac{1}{2}x + 7$.

55. Notice the graph of the function shown to the right.

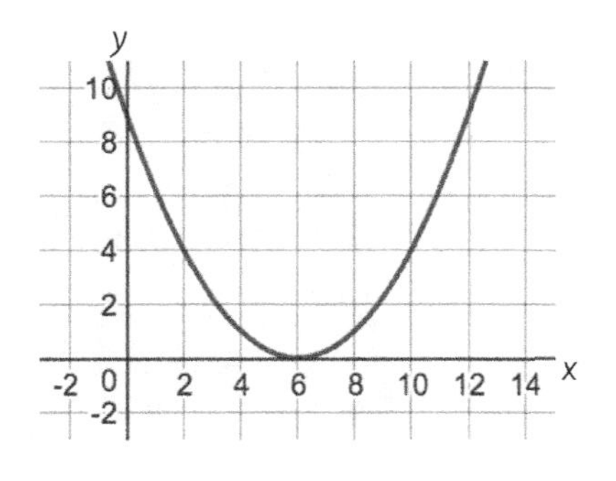

a. How many x–intercepts does this function have?

The shape of this function is known as a parabola (pronounced puh-RA-buh-luh). If you tossed a small rock into the parabola, notice where it would settle: at the bottom. The location where the rock would settle is called the vertex of the parabola.

b. Estimate the coordinates of the vertex of this parabola.

c. Estimate the y-intercept of this parabola.

56. Determine the x-intercepts of the function $f(x)=\frac{1}{4}x^2-3x+9$. Use the method of Completing the Square, since it is the most recent strategy that you have learned. The graph of this function is shown in the previous scenario, which should help you see if your x–intercepts are accurate.

57. If the parabola in the previous scenario is shifted down 1 unit, how many x-intercepts will it have?

58. If the parabola in the previous scenario is shifted up 2 units, how many x-intercepts will it have?

59. A function is shown below. Find and plot the x-intercepts of the function in the graph shown.

$y=x^2-6x+13$

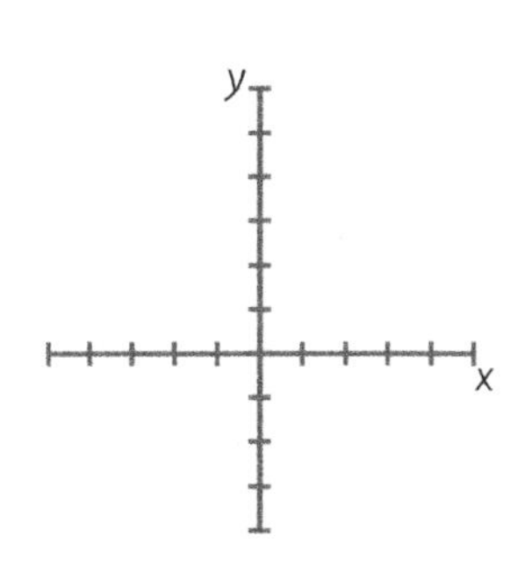

Section 6

SOLVING QUADRATIC EQUATIONS WITH THE QUADRATIC FORMULA

60. The Quadratic Formula seems complicated at first, but it will become more familiar as you use it to solve quadratic equations. For now, write out The Quadratic Formula.

61. Simplify the following expressions.

a. $\dfrac{-(9)\pm\sqrt{(9)^2-4(-2)(5)}}{2(-2)}$

b. $\dfrac{-(-8)\pm\sqrt{(-8)^2-4(2)(8)}}{2(2)}$

62. What is the value of the expression $\dfrac{-b\pm\sqrt{b^2-4ac}}{2a}$ if a = 3, b = -5 and c = 2?

63. What is the value of the expression $\dfrac{-b\pm\sqrt{b^2-4ac}}{2a}$ if a = -2, b = 2 and c = -6?

64. If you wanted to solve each of the following equations using the Quadratic Formula, what numbers would you use for a, b, and c?

a. $3x^2-7x+2=0$

b. $6x^2-x=1$

c. $x^2+4x-2=9x-11$

65. Earlier, you found the solution to the equation $x^2+4x+1=0$. Use the Quadratic Formula to solve the equation and confirm that you arrive at the same solution of $x=-2\pm\sqrt{3}$.

66. Now use the Quadratic Formula to find the solution to the equation $x^2+4x+10=0$. Notice the subtle difference between this equation and the previous equation.

67. Consider the equation $2x^2 - 6x + 3 = 0$.

a. Use the Quadratic Formula to confirm that the solution is $x = \frac{3 \pm \sqrt{3}}{2}$.

b. Use a calculator to determine the value of $\frac{3 \pm \sqrt{3}}{2}$ rounded to the nearest tenth.

68. In the equation $2x^2 - 6x + 3 = 0$, if you move the three terms to the other side of the equation, it becomes $0 = -2x^2 + 6x - 3$, but the solution must still be the same. Without changing the numbers, use the Quadratic Formula to solve the equation $0 = -2x^2 + 6x - 3$.

69. Determine the x-intercepts of the parabola modeled by the function $f(x) = -\frac{1}{3}x^2 + 2x + 4$.

a. Determine the exact value of the x-intercepts.

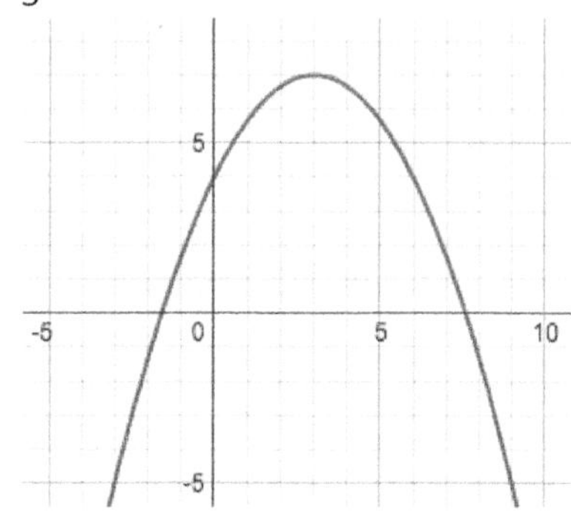

b. Use a calculator to approximate the value of the x-intercepts to the nearest tenth. Refer to the graph shown to the right to verify that your results seem accurate.

c. Use the graph to estimate the coordinates of the vertex of this parabola.

70. Locate the exact x-intercepts of the function $g(x)=-\frac{1}{3}x^2+2x-4$. Use the Quadratic Formula, since it is the most recent strategy that you have learned.

71. ★One of the solutions to the equation $x^2-2x-10=0$ is $x=1+\sqrt{11}$. Plug this value into the original equation to show that it makes the original equation true.

72. ★The previous equation has 2 solutions. Without doing any calculations, what is the other solution?

Section 7
REVIEW

73. Solve each equation.

a. $x^2 = 44$

b. $x^2 = -\frac{4}{25}$

74. Solve each equation by factoring.

a. $4x^2 - 4x + 1 = 0$

b. $x^2 = 5x$

75. Round the expression $\frac{-7 \pm 2\sqrt{5}}{5}$ to the nearest tenth.

76. Complete the trinomial to make it a perfect square trinomial.

a. $x^2 - 12x + \_\_\_\_$

b. $x^2 + \frac{1}{2}x + \_\_\_\_$

77. Factor each trinomial in the previous scenario.

78. Without changing it, can this equation be solved using the Quadratic Formula? If it can, what values would you use for a, b, and c? Do not solve the equation.

$-x^2 + 11x = 2$

79. Without changing it, can this equation be solved using the Quadratic Formula? If it can, what values would you use for a, b, and c? Do not solve the equation.

$\frac{1}{5}x^2 - 3x = 0$

80. Use the Quadratic Formula to solve the equation $-\frac{1}{5}x^2+3x=0$.

81. Each equation has two solutions. What is the average of the two solutions?

a. $(x-2)^2=81$

b. $(x+8)^2=7$

82. Given the equation below, what is the value of y if x is -5?

$y=-2x^2+15$

Section 8

THE VERTEX OF A PARABOLA

83. Draw a graph for each function below, using a T-chart. Include at least 6 points in your graph.

a. $y = x^2 - 2$

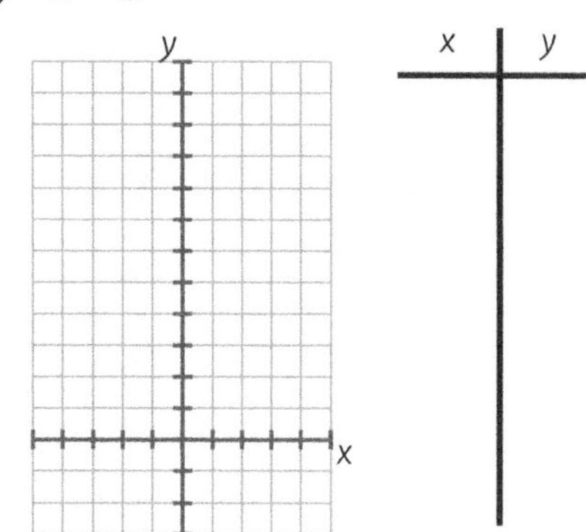

b. $y = -x^2 + 1$

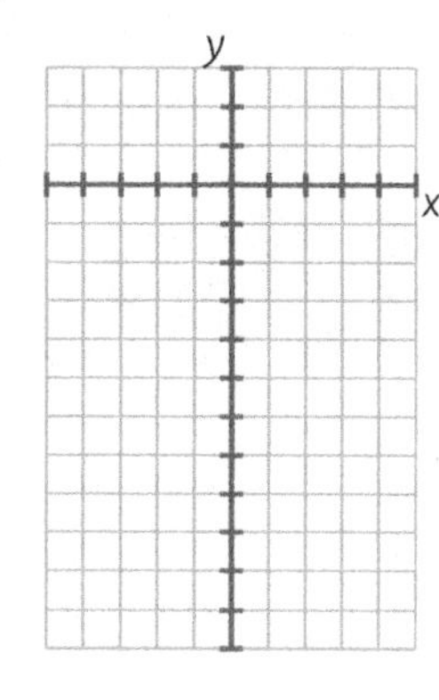

84. Identify the coordinates of the vertex of each parabola above.

85. Look at each of the parabolas below. Estimate the coordinates of the x-intercepts. Estimate the coordinates of the vertex.

a.

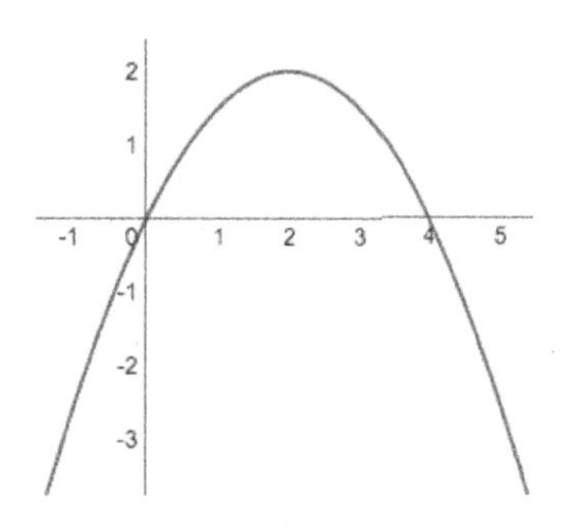

b.

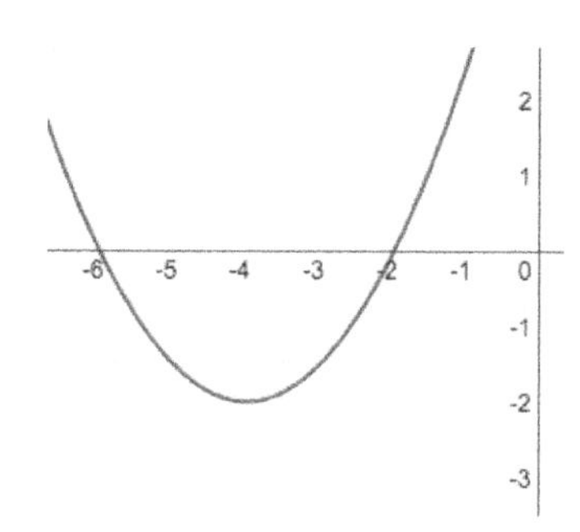

86. In the previous scenario, compare the location of the vertex and the location of the x-intercepts.

87. If the x-intercepts of a parabola are (2, 0) and (8, 0), what is the x-value of the vertex?

88. On the parabola shown, find the point with the greatest y-value. What is the x-value at this point?

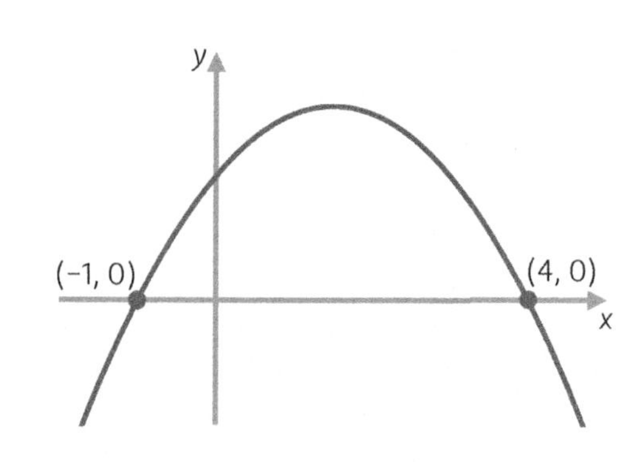

89. What is the x-value of the vertex of a parabola if its x-intercepts are located at the points shown?

a. (7, 0) and (−3, 0)

b. (−4, 0) and (5, 0)

c. (x, 0) and (5, 0)

d. $\left(4+\sqrt{7},0\right)$ and $\left(4-\sqrt{7},0\right)$

90. If the vertex of a parabola is located at (3, 2) and one of the x-intercepts is located at (−2, 0), where is the other x-intercept located? If it helps, draw a simple sketch.

91. Identify the coordinates of both of the x-intercepts of the parabola shown in the graph.

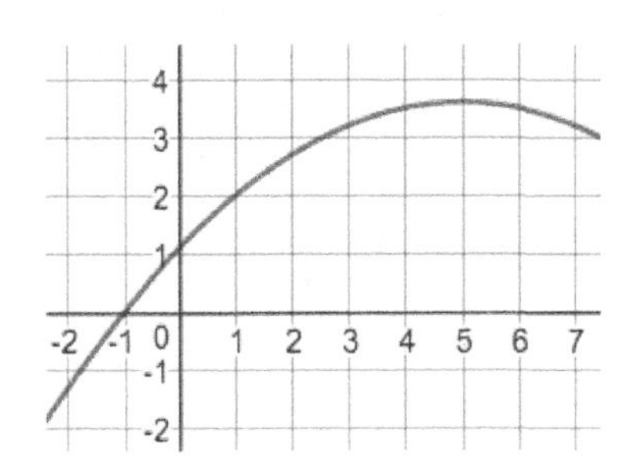

One feature that makes a parabola special is its symmetry. Each parabola can be cut in half along its line of symmetry, which passes through the vertex. As a result, knowing the location of the vertex helps determine other points along the parabola.

92. One way to find the vertex is to use a parabola's symmetry. First, locate the x-intercepts. Then, find the x-value halfway between them. For each function below, determine the x-value of the vertex.

a. $y=x^2-4$

b. $f\left(x\right)=x^2+5x-14$

93. ★Find the x-value of the vertex of the parabola modeled by the function $g(x)=\frac{1}{3}x^2-2x+1$.

94. Write the number that is halfway between the two numbers shown.

a. $\sqrt{5}$ and $-\sqrt{5}$

b. $9+\sqrt{5}$ and $9-\sqrt{5}$

c. $\frac{2}{7}+\frac{\sqrt{5}}{7}$ and $\frac{2}{7}-\frac{\sqrt{5}}{7}$

95. Write the number that is halfway between the two numbers shown.

a. $7\pm\sqrt{2}$

b. $F\pm\sqrt{G}$

c. $\frac{y}{3}\pm\frac{4}{3}$

d. $-\frac{b}{2a}\pm\sqrt{x}$

96. For a quadratic function in the form $y=ax^2+bx+c$, you can locate the x-intercepts by solving the equation $0=ax^2+bx+c$. If you use the Quadratic Formula to solve this equation, the solution is $x=\frac{-b\pm\sqrt{b^2-4ac}}{2a}$. Separate this fraction to make it $x=\frac{-b}{2a}\pm\frac{\sqrt{b^2-4ac}}{2a}$.

a. What number is halfway between $x=\frac{-b}{2a}\pm\frac{\sqrt{b^2-4ac}}{2a}$?

b. Since the vertex of a parabola is always located halfway between the x-intercepts, the vertex will always have an x-value of _______.

97. If a = 4 and b = -2, what is the value of $\frac{-b}{2a}$?

98. What is the value of $\frac{-b}{2a}$ if a and b are assigned the values shown below?

b. a = -1 and b = -10

b. a = $-\frac{1}{5}$ and b = -2

99. For each function below, determine the x-value of the vertex by using expression $x=\frac{-b}{2a}$.

a. $f(x)=x^2-4$

b. $g(x)=x^2+5x-14$

★c. $h(x)=\frac{1}{3}x^2-2x+1$

100. If you only know the x-value of the vertex, you only have half of the location of that point. You still need the y-value before you can plot the vertex point.

a. If you have the equation for a parabola, how can you find the y-value of its vertex, if you only know the x-value?

b. For the parabola $y = x^2 + 6x - 7$, the vertex is located at (−3, ___). Fill in the blank.

101. Find the coordinates of the vertex for each quadratic function shown below. Use $x = \frac{-b}{2a}$ to find the x-value. Then use the function to find the y-value.

a. $a(x) = -x^2 + 6x + 3$

b. $b(x) = 4 - x^2$

c. $c(x) = 3x^2 - 12x + 6$

102. ★Locate the coordinates of the vertex of the parabola modeled by the function $d(x) = 1 - 2x - \frac{1}{5}x^2$.

103. Part of a parabola has already been graphed for you. Use what you have learned about the symmetry of a parabola to complete each graph.

a.

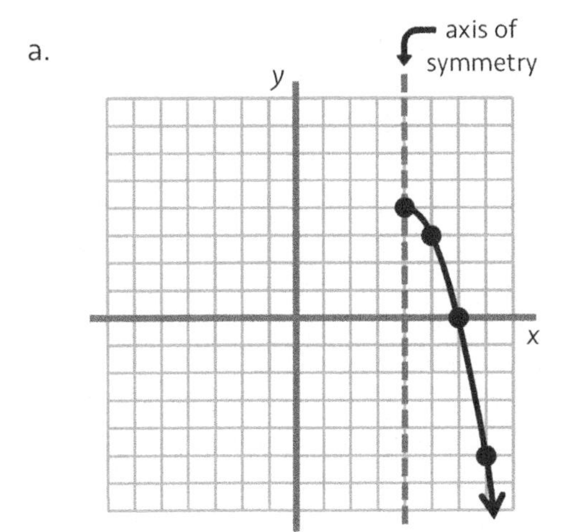

b.

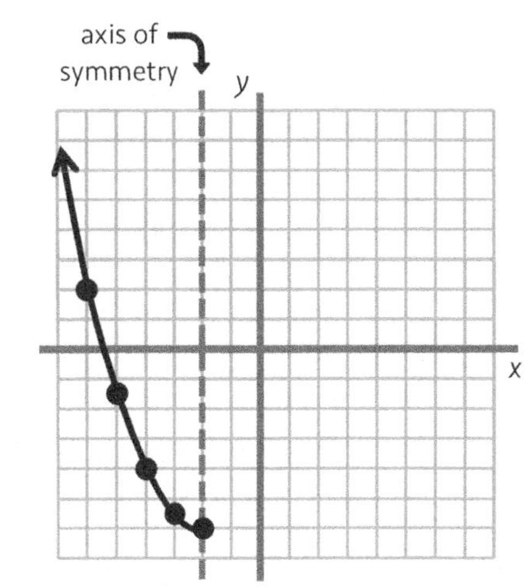

104. In the graphs in the previous scenario, the axis of symmetry is marked for you. This line is drawn to help you use the symmetry of the parabola to find more points. It is a dashed line to show that it is not actually a visible line. Draw the axis of symmetry on the graph of each parabola below.

a.

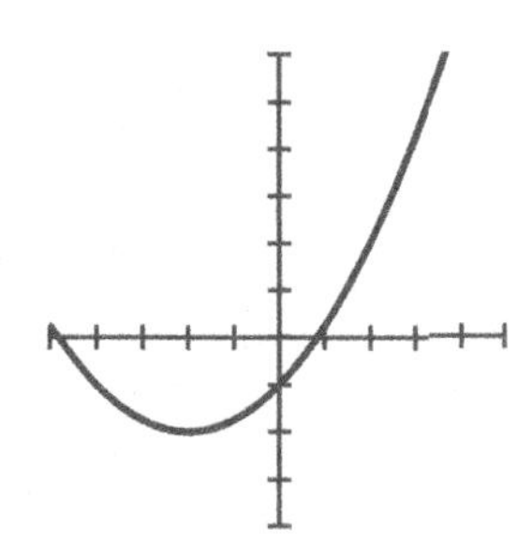

b.

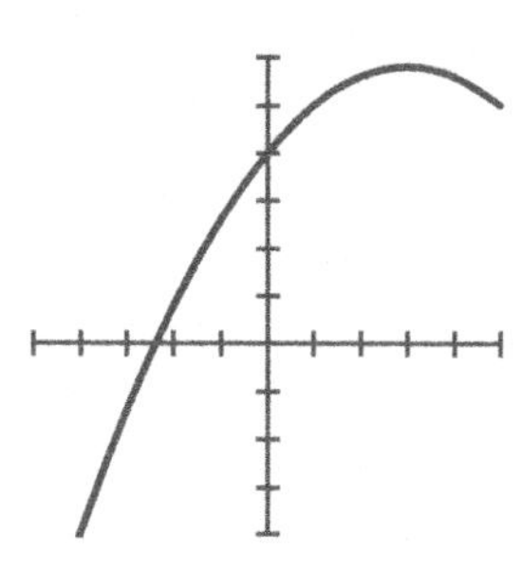

105. Write the equation for the line of symmetry that you drew in each graph in the previous scenario.

106. What is the equation for the axis of symmetry of the graph of $y=-2x^2-12x+5$?

107. What is the y-intercept of the graph of $y=\frac{1}{2}x-7$?

108. What is the y-intercept of the graph of $y=3x^2-2x+11$?

109. Solve for x in each equation.

a. $x=20-2(-10)^2$

b. $(x-4)^2=9$

c. $x^2-2x=0$

Section 9

GRAPHING A PARABOLA

110. Graph the function $y = x^2 - 4x + 4$ by finding points as directed.

a. Find the vertex.

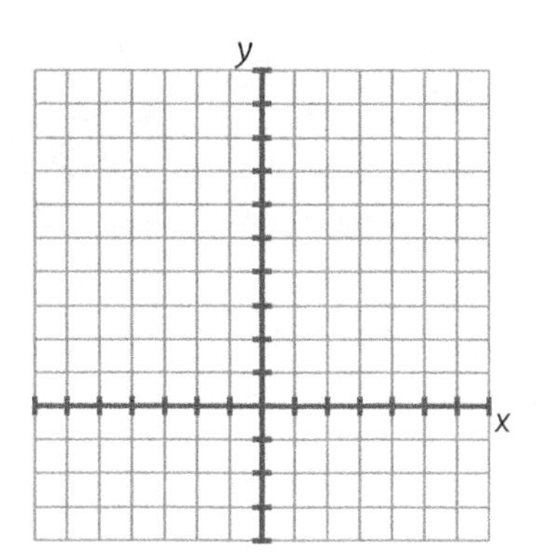

b. Find the y-intercept by letting $x = 0$.

c. Find the x-intercepts by letting $y = 0$.

d. Plot other points by plugging in other x-values.

e. Plot the points that you have found on the graph shown.

f. Draw a smooth parabola through the points.

111. Graph the function $y = x^2 + 4x$ by plotting at least 6 points.

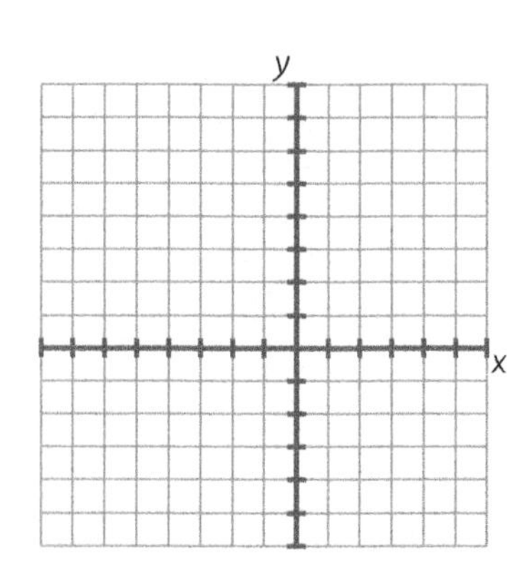

112. ★Graph each parabola by using the strategy listed in the previous scenario. Plot at least six points.

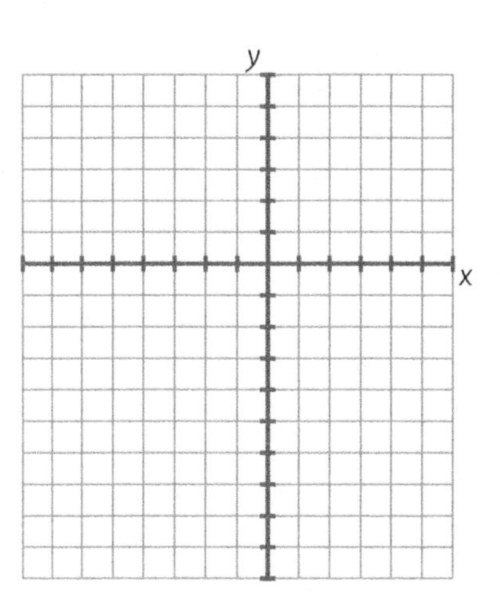

$$g(x) = -\frac{1}{2}x^2 - 2x + 1$$

113. ★Graph the parabola modeled by the function $h(x) = \frac{1}{4}x^2 - x - 3$. Plot at least six points.

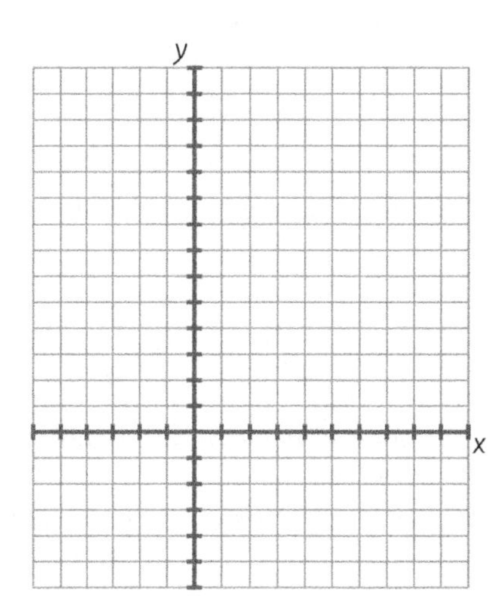

114. Identify the vertex and x-intercepts of the parabola modeled by $f(x) = -x^2 + 6x - 10$.

115. Make the equation match the graph below it by filling in the blank.

a. $y=\frac{1}{4}x^2+\_\_\_\_x-1$

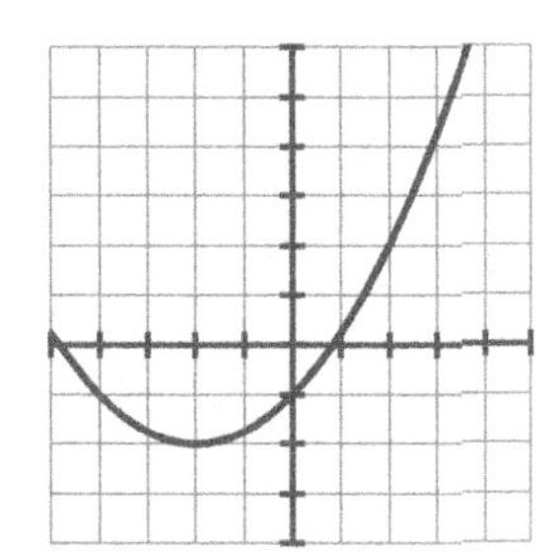

★b. $y=-\frac{1}{5}x^2+\_\_\_\_x+4$

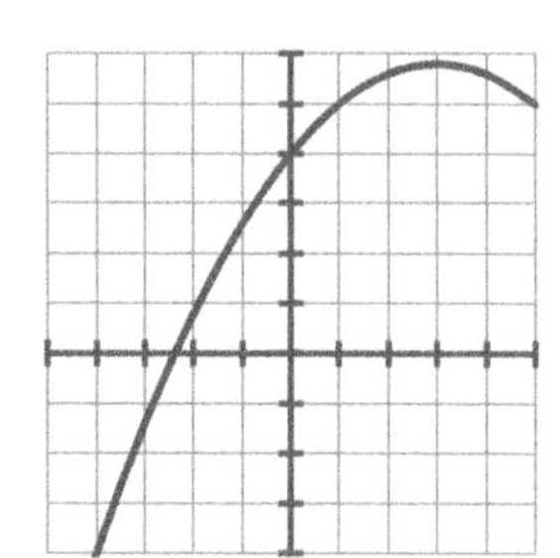

When you graph a quadratic function, the shape of the parabola will roughly resemble either ∪ or ∩. If its shape is ∪, it opens upward because, like a cup, the opening is up. If its shape is ∩, it opens downward. Imagine tipping a cup upside down when it is full of water. The water would fall downward to the floor. If you haven't realized it yet, there is a simple way to determine whether a parabola will open upward or downward. Go back through all of the scenarios that contain a parabola and its equation and see if you can figure out this relationship.

116. How can you use a parabola's equation to determine if it opens upward or downward?

117. Look at each function below and state whether the parabola will open upward or downward.

a. $f(x)=-\frac{1}{2}x^2+\frac{3}{2}x$

b. $A(t)=t^2+8t-2$

c. $P(n)=3.3-7.9n^2$

118. Draw a very basic sketch of a parabola that has each of the following characteristics.

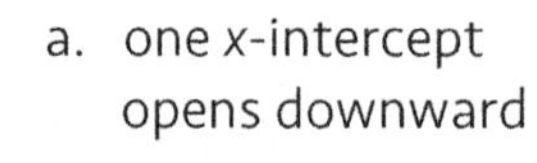

a. one x-intercept
opens downward

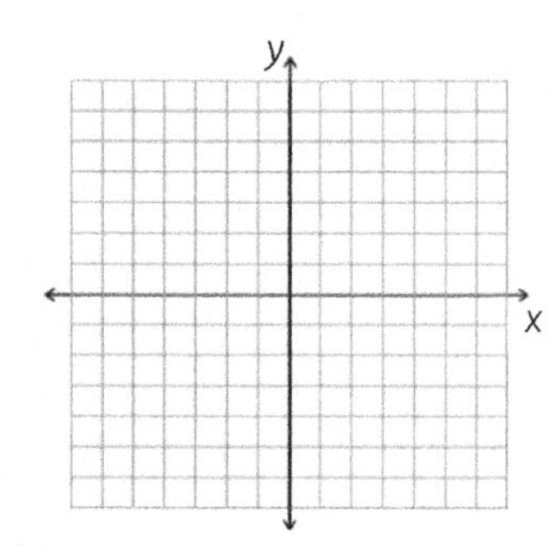

b. no x-intercepts
opens upward

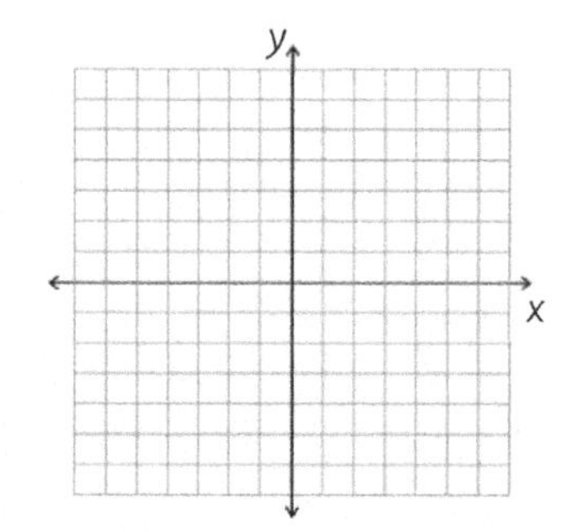

c. two x-intercepts
opens upward

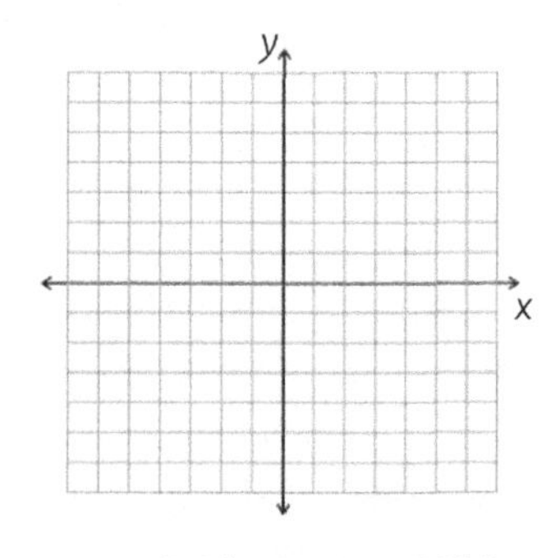

119. Which of the following functions could represent the parabola shown in the graph? Circle all possible matches. For each function that could not represent the parabola shown, briefly explain why the function could not match the graph.

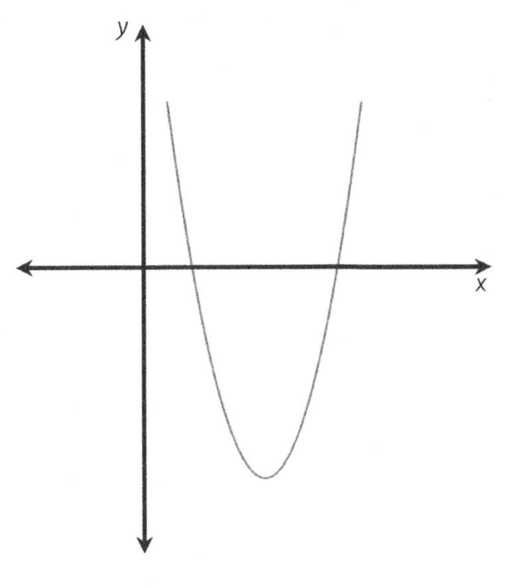

a. $y=-x^2+2x-5$

b. $y=x^2-3x-1$

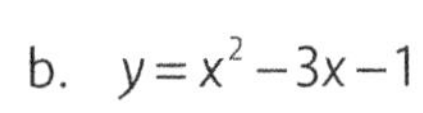

c. $y=x^2+2x+8$

d. $y=(x-2)(x-8)$

e. $y=x(x-8)$

120. Graph the parabola by plotting the vertex and at least six other points.

$H(t)=-2t^2+8t+1$

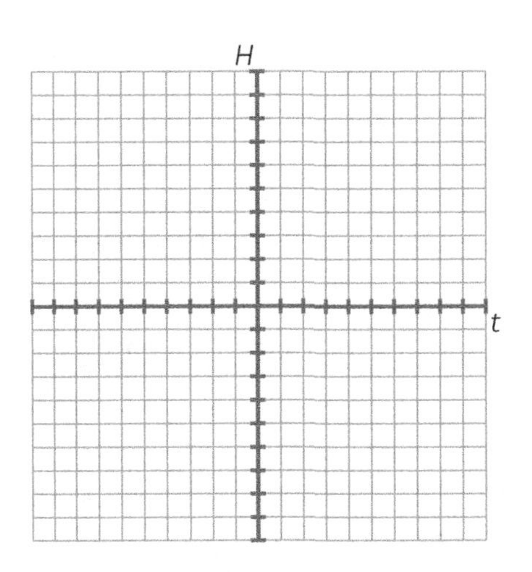

121. Which parabola has x-intercepts that are farther apart? Answer this question without graphing or doing any numerical calculations.

Parabola #1: $y=3x^2-17x-41$

Parabola #2: $y=4x^2-17x-41$

122. Go to the website www.desmos.com/calculator/ and graph the following functions on the same graph. What do you notice?

$y=x^2+2x-4$ $\qquad$ $y=-4x^2-8x+16$ $\qquad$ $y=\frac{1}{2}x^2+x-2$

123. Why do each of the parabolas in the previous scenario have the same x-intercepts?

124. ★Unlike lines, parabolas do not have a constant slope. Consider each parabola shown below.

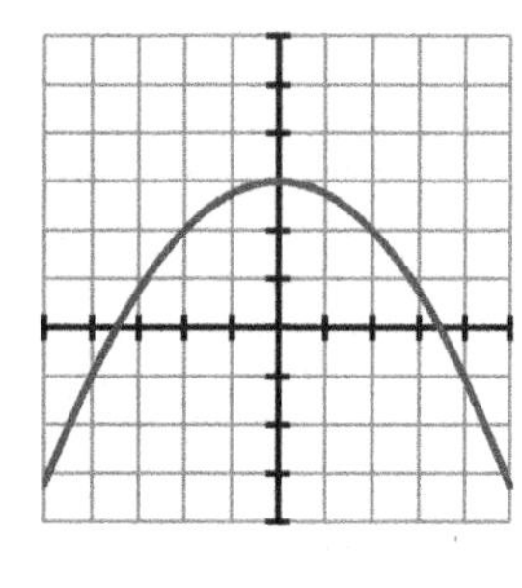

$f(x)$

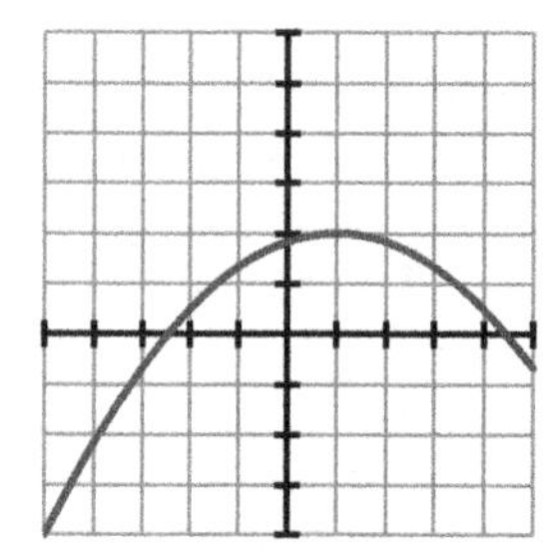

$g(x)$

a. Is the slope steeper at $f(-2)$ or $g(-4)$?

b. Estimate the numerical value of the slope of $f(x)$ at x = 0.

125. In the previous scenario, if the parabola continues without ending beyond what is shown in the graph, over what interval is the graph of $g(x)$ decreasing? Write this as an inequality.

126. In the previous scenario, which value is greater?

a. $f(2)$ or $g(0)$? $\qquad$ b. $f(-5)$ or $g(-5)$? $\qquad$ c. $f(3)$ or $g(3.5)$

Section 10

SCENARIOS THAT INVOLVE QUADRATIC FUNCTIONS

127. Before a softball game, two players toss a ball back and forth in the outfield. Each time the ball is thrown through the air, it follows a path modeled by the function $H(t) = -16t^2 + 48t + 6$, where H is the height above the ground, in feet, and t is how long the ball has been in the air, in seconds.

a. What is the ball's height above the ground at the moment it is thrown?

b. What is the maximum height of the ball as it travels through the air?

128. In the previous scenario, the ball is thrown by one player and caught by the other player when it is 6 feet above the ground. How long is the ball in the air?

129. For how many seconds is the ball higher than 38 feet above the ground as it travels through the air in the previous scenario?

130. For how many seconds is the ball higher than 45 feet above the ground as it travels through the air in the previous scenario?

131. ★Janet and Brett launch water balloons from separate slingshots. The paths of the water balloons as they travel through the air are modeled by the functions shown below. $J(x)$ and $B(x)$ represent the height of each water balloon above the ground, while x represents the horizontal distance traveled by the water balloon. The values of x, $J(x)$ and $B(x)$ are all measured in feet.

Janet	Brett
$J(x) = -0.04x^2 + 1.6x + 1$	$B(x) = -0.05x^2 + 1.6x + 1$

Whose water balloon travels farther? Answer this question without doing any numerical calculations.

132. Use the graphing calculator on www.desmos.com to graph the functions in the previous scenario. Use the graphs to help you answer the question in the previous scenario.

133. In the previous scenario, which person's water balloon reached the greatest height as it traveled through the air? By how many feet was its maximum height greater the maximum height of the other water balloon?

134. If you have a wire that is 24 inches long, you can make rectangles of many different sizes.

a. List the dimensions of 3 different rectangles that you could make with 24 in. of wire.

b. The area of a rectangle made with 24 in. of wire is a function of the width of the rectangle and it can be modeled by the equation $A(w) = w(12 - w)$, where A is the area if the width has a measurement of w. Write the area function in the form $A(w) = \_\_ w^2 + \_\_ w + \_\_$.

c. What is the maximum area of a rectangle made with 24 inches of wire?

135. If you have only 32 inches of wire, and you use all of the wire to form a rectangle with the largest possible area, what is the area of that rectangle?

136. At the end of a game, with the score tied, a kicker lines up to attempt a 55-yard field goal. The flight of the ball after it is kicked follows a parabola modeled by the function $H(x) = -0.01x^2 + 1.7x$, where H is the height of the ball in feet after it has traveled horizontally x feet.

a. When the ball is kicked, it is on the ground. How many feet does the ball travel horizontally before it reaches a height of 30 feet above the ground?

b. How many feet does the ball travel horizontally while its height is above 30 feet?

137. ★In the previous scenario, will the ball go over the crossbar if the bar is 10 feet above the ground?

138. During the next game, a kicker once again attempts a game-winning field goal. This time, the flight of the ball is modeled by the function $H(x) = -0.01x^2 + 1.525x$, where H is the height of the ball in feet after it has traveled horizontally x feet. If the attempt misses because the ball hits the crossbar instead of going over it, how long was the field goal attempt (how far did the ball travel horizontally)? Approximate this distance by rounding to the nearest tenth.

139. Ilana has been training all season to break the girls national high school shot put record, which is 56.7 feet. At the last meet of the season, she achieves her best throw. The path of the shot as it travels through the air can be modeled by the function $H=-0.02d^2+1.02d+7$, where H is the height of the shot, in feet, and d is the distance the shot has traveled horizontally, also measured in feet. Does her final throw break the record?

140. ★After the shot put hits the ground in the previous scenario, what is the distance, in inches, between Ilana's throw and the national record?

141. ★What is the horizontal distance between Ilana's release point and the shot when the shot put reaches its maximum height in the previous scenario?

142. Vera owns a bagel shop. She has studied how customers respond to price changes and if she makes the bagels too cheap, the demand is high but her company's profits (money earned after expenses are subtracted) are small. As she increases the price, the profits seem to increase as well. She estimates that her monthly profit can be modeled by the function $P=-25d^2+75d-36$, where P is the monthly profit, in thousands of dollars, if she sells each bagel for d dollars.

a. Can she keep raising prices and expect her company's profits to continue to increase?

b. Does her model suggest that it would be better to sell her bagels for \$1.20 or \$1.60?

c. What happens when Vera raises the price from \$1.60 to \$2.00 per bagel?

Section 11

GRAPHING QUADRATIC INEQUALITIES

143. Bring your memory back to a previous topic: graphing linear inequalities. Graph the following inequalities.

a. $y<4$

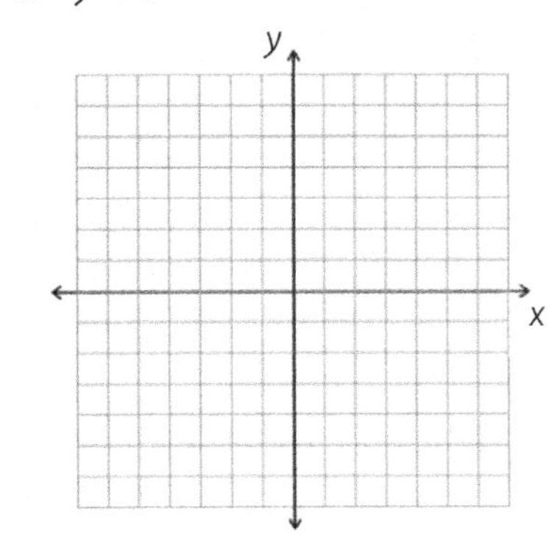

b. $6x+2y\geq-4$

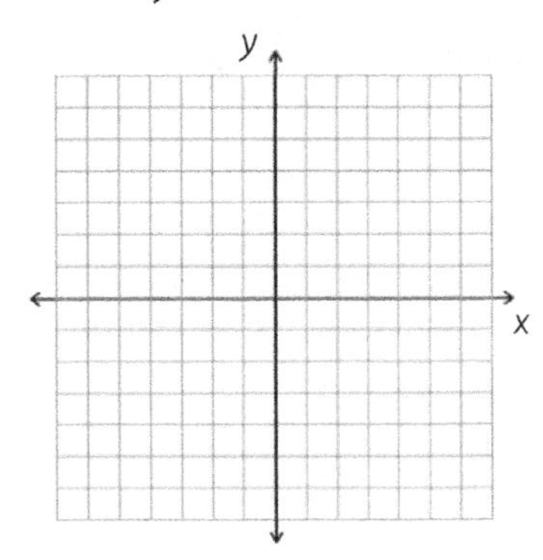

144. Graph the following quadratic inequality.

$y\geq x^2$

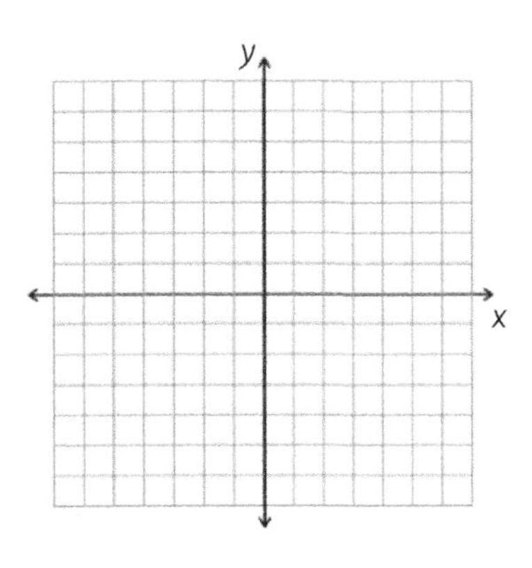

145. Graph the following quadratic inequalities.

$y<-x^2+2x+3$

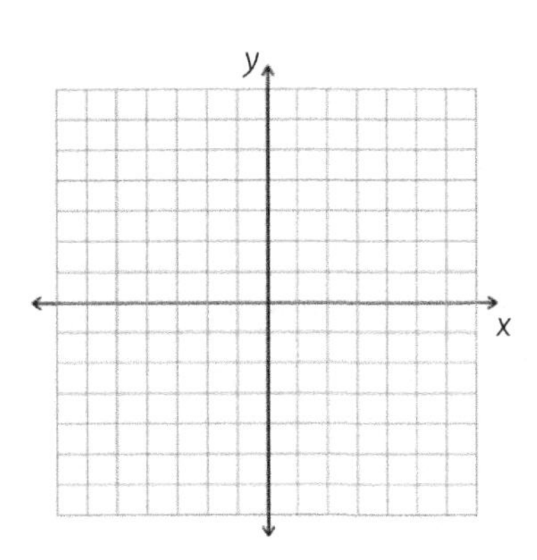

Section 12
CUMULATIVE REVIEW

146. Consider the two lines shown to the right.

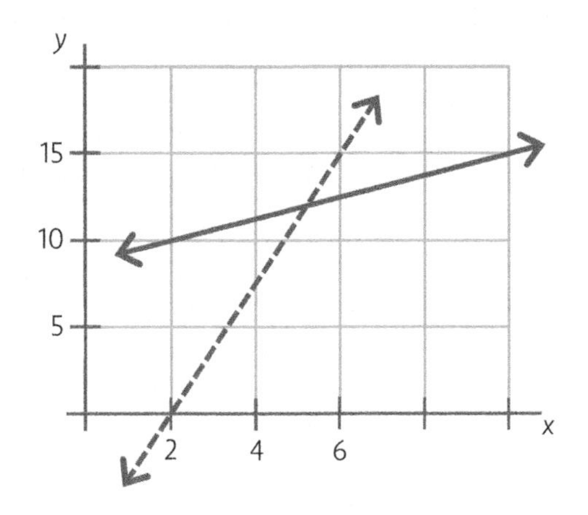

a. Identify the equation of the dashed line and write the equation in Slope-Intercept Form.

b. Write the equation of the solid line in Point-Slope Form.

★c. Determine the exact location of the point where the two lines intersect.

147. The chart to the right shows a small data set.

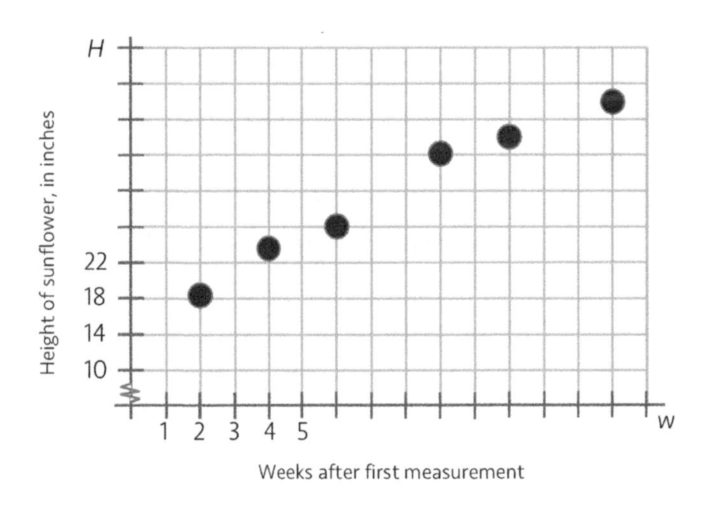

a. Draw an approximate trend line through the scatter plot shown.

b. Find the equation of your trend line to show the relationship between H, the total height, and w, the weeks after the first measurement.

148. Use your work in the previous scenario to answer the following questions.

a. What was the average increase in the height per week over the time period shown?

b. Estimate the height of the sunflower when it was first measured.

149. Consider the equation $\frac{18}{y}=4+2x$.

a. Find x in terms of y. (This means to isolate x.) Write the expression that contains y as a fully simplified fraction.

★b. Find y in terms of x.

150. Which parabola has x-intercepts that are closer together? Explain your choice without performing numerical calculations.

Parabola #1

$y=1.4x^2+3x-4$

Parabola #2

$y=1.5x^2+3x-4$

151. Factor each polynomial.

a. $-12x^2+75$

b. $4x^2+26x+42$

152. Jan's hourly wage was increased from \$20 to \$22 per hour. Fred changed jobs and his hourly wage was reduced from \$22 to \$20. Whose wage changed by a greater percent? Explain your conclusion.

Section 13

ANSWER KEY

1.	a. $\frac{3}{8}$ b. $\frac{8}{3}$ c. $\frac{3}{2}$ d. $\frac{2}{3}$
2.	a. 81 b. $-2\cdot 4\rightarrow -8$ c. $1-49\rightarrow -48$
3.	a. −3 b. Two x-values, −1 and 3. c. $f(-4)>f(-3)\rightarrow f(-4)=5;\ f(-3)=3$
4.	a. $0<x<6$ b. $-5<x<0$
5.	a. $y=-\frac{1}{4}$ b. $y=\frac{7}{4}$
6.	*f* of *x* equals three *x* squared plus five *x* minus one.
7.	a. −1 b. 21 c. 1
8.	a. $\frac{1}{\sqrt{7}}\cdot\frac{\sqrt{7}}{\sqrt{7}}=\frac{\sqrt{7}}{7}$ b. $\frac{\sqrt{7}}{\sqrt{2}}\cdot\frac{\sqrt{2}}{\sqrt{2}}=\frac{\sqrt{14}}{2}$ c. $\frac{3\sqrt{5}}{\sqrt{2}}\cdot\frac{\sqrt{2}}{\sqrt{2}}=\frac{3\sqrt{10}}{2}$
9.	In the function, replace *d* with 56.7 to find that $H\approx -0.6$. This confirms that her shot does not break the record. Since the height is negative, the shot would be under the ground. This shows that the shot hit the ground *before* d = 56.7.
10.	$H(0)=-0.02(0)^2+(0)+7\rightarrow H(0)=7$ At the moment the shot put leaves Ilana's hand, it is 7 feet above the ground.
11.	a. −6 b. −12 c. 3
12.	a. $(x-4)(x-7)$ b. $(2x-3)(x+4)$ c. $(3x-2)(3x-2)$
13.	a. x = −7, −2 b. $x=\pm 5$ c. x = 1
14.	(2,0) → solve 8 − 4x = 0
15.	(3,0), (−3,0) → solve $2x^2-18=0$
16.	a. 9 b. 4 c. 12
17.	a. $3\sqrt{5}$ b. $2\sqrt{10}$ c. $5\sqrt{2}$
18.	a. $\frac{4}{5}$ b. $\frac{\sqrt{6}}{5}$ c. $\frac{2\sqrt{6}}{5}$ d. 0 e. $\sqrt{2}$
19.	i
20.	imaginary (as a side note, numbers that do have square roots have been given the name real numbers in response to the naming of imaginary numbers)

21.	$5i$
22.	a. 10 b. 6 c. $\frac{3}{5}$ d. $-\frac{1}{36}$
23.	a. $2i\sqrt{5}$ b. $2i\sqrt{10}$ c. $4i\sqrt{5}$
24.	a. $\frac{1}{2}$ b. $\frac{\sqrt{11}}{4}$ c. $\frac{\sqrt{6}}{3}$
25.	a. $\frac{1}{2}i$ b. $\frac{2i\sqrt{6}}{5}$ c. $\frac{2i}{\sqrt{7}}\rightarrow\frac{2i\sqrt{7}}{7}$
26.	$i^2=-1$
27.	a. $9i^2\rightarrow 9(-1)\rightarrow -9$ b. $9i^2\rightarrow 9(-1)\rightarrow -9$ c. $36i^2\rightarrow 36(-1)\rightarrow -36$ d. −5
28.	Each side is 5 in. The perimeter is 20 in.
29.	a. Each side is $\sqrt{200}$ or $10\sqrt{2}$ ft. b. $10\sqrt{2}\approx 14.14\rightarrow$ Perimeter: 56.6ft
30.	a. $w=\pm 3$ b. $x=\pm\sqrt{11}$ c. $y=\pm\frac{\sqrt{3}}{2}$
31.	Squaring a positive number gives the same result as squaring a negative number, so there are 2 possible solutions to an equation with a structure $x^2=$ ____ .
32.	a. $x=4\pm 1\rightarrow x=3,\ 5$ b. $x=1\pm\sqrt{13}$ c. $x=-\frac{1}{3}\pm\frac{\sqrt{2}}{3}$
33.	$x=-11\pm\sqrt{R}$
34.	a. $x=3\pm\sqrt{2}$ b. $x=-7\pm\frac{1}{4}i$ c. $x=\frac{4}{3}\pm\frac{4}{3}\rightarrow x=\frac{8}{3},\ 0$
35.	a. $x=\pm i$ b. $x=\pm i\sqrt{10}$ c. $x=\pm\frac{4}{5}i$
36.	a. 6 b. 22, 121
37.	a. 7, 49 b. $\frac{1}{2},\ \frac{1}{4}$
38.	a. $B=2A$ b. $C=\left(\frac{B}{2}\right)^2$
39.	a. 25 b. 4 c. 100

40.	a. $(x+5)^2$ b. $(x-2)^2$ c. $(x-10)^2$
41.	a. 144 b. $\frac{25}{4}$ c. $\frac{49}{4}$
42.	a. $(x+12)^2$ b. $\left(x+\frac{5}{2}\right)^2$ c. $\left(x-\frac{7}{2}\right)^2$
43.	Add 2 to both sides. (You could also 51 to both sides, but that might seem confusing at this point.)
44.	$x^2-14x \quad =15 \rightarrow x^2-14x \ \underline{+49}=15 \ \underline{+49}$ $(x-7)^2=64 \rightarrow x-7=\pm 8 \rightarrow x=15 \text{ or } -1$
45.	Compute $\frac{12}{2}$ to get 6. Compute 6^2 to get 36. Add 36 to both sides of the equation.
46.	$x^2+12x+36=40 \rightarrow (x+6)^2=40$ $\rightarrow x+6=\pm\sqrt{40} \rightarrow x=-6\pm 2\sqrt{10}$
47.	$x^2+2x \quad =15 \rightarrow x^2+2x \ \underline{+1}=15 \ \underline{+1}$ $(x+1)^2=16 \rightarrow x+1=\pm 4 \rightarrow x=3 \text{ or } -5$
48.	$x^2+4x \ \underline{+4}=-1 \ \underline{+4} \rightarrow (x+2)^2=3$ $x+2=\pm\sqrt{3} \rightarrow x=-2\pm\sqrt{3}$
49.	$x = -0.27, -3.73$
50.	$x=6\pm 2\sqrt{7}$
51.	divide both sides by −4 to make the x^2 term have a coefficient of "1"
52.	a. $x=3\pm 2\sqrt{3}$ b. $x=-5\pm 3\sqrt{2}$
53.	a. $x = 6.46, -0.46$ b. $x = -0.76, -9.24$
54.	a. Let $y = 0$. Solve for x. b. $(-14,0)$
55.	a. One b. $(6,0)$ c. $(0,9)$
56.	$\frac{1}{4}x^2-3x+9=0 \rightarrow 4\cdot\left(\frac{1}{4}x^2-3x+9\right)=(0)\cdot 4$ $x^2-12x+36=0 \rightarrow (x-6)^2=0 \rightarrow x=6$ The x-intercept is $(6,0)$.
57.	The vertex will be below the x-axis, so the parabola will have two x-intercepts.
58.	The vertex will be above the x-axis, so the parabola will have no x-intercepts. It will not cross the x-axis.
59.	The x-intercepts are $3 + 2i$ and $3 - 2i$. This shows that the function does not cross the x-axis. The x-intercepts contain imaginary values. Since they are imaginary, they cannot be plotted on a graph.

60.	$x=\frac{-b\pm\sqrt{b^2-4ac}}{2a}$
61.	a. $\frac{-9\pm 11}{-4} \rightarrow -\frac{1}{2}, 5$ b. $\frac{8\pm 0}{4} \rightarrow 2$
62.	$\frac{-(-5)\pm\sqrt{(-5)^2-4(3)(2)}}{2(3)} \rightarrow \frac{5\pm\sqrt{1}}{6} \rightarrow$ $\frac{5\pm 1}{6} \rightarrow \frac{6}{6} \text{ or } \frac{4}{6} \rightarrow 1 \text{ or } \frac{2}{3}$
63.	$\frac{-(2)\pm\sqrt{(2)^2-4(-2)(-6)}}{2(-2)} \rightarrow \frac{-2\pm\sqrt{-44}}{-4} \rightarrow$ $\frac{-2\pm 2i\sqrt{11}}{-4} \rightarrow \frac{-1\pm i\sqrt{11}}{-2} \text{ or } \frac{1\pm i\sqrt{11}}{2}$
64.	a. $a = 3, \ b = -7, \ c = 2$ b. $a = 6, \ b = -1, \ c = -1$ c. $a = 1, \ b = -5, \ c = 9$
65.	-
66.	$x=-2\pm i\sqrt{6}$
67.	a. - b. 2.4, 0.6
68.	$x=\frac{3\pm\sqrt{3}}{2}$
69.	a. $x=3\pm\sqrt{21}$ b. 7.6, −1.6 c. (3,7)
70.	Solve $0=-\frac{1}{3}x^2+2x-4 \rightarrow x=3\pm i\sqrt{3}$. The function does not have any real x-intercepts because there are imaginary numbers in the solution. This is a parabola that does not cross the x-axis.
71.	$\left(1+\sqrt{11}\right)^2-2\left(1+\sqrt{11}\right)-10=0$ $1+2\sqrt{11}+11-2-2\sqrt{11}-10=0 \rightarrow 0=0$
72.	$x=1-\sqrt{11}$
73.	a. $x=\pm 2\sqrt{11}$ b. $x=\pm\frac{2}{5}i \text{ or } \pm\frac{2i}{5}$
74.	a. $(2x-1)(2x-1)=0 \rightarrow x=\frac{1}{2}$ b. $x^2-5x=0 \rightarrow x(x-5)=0 \rightarrow x=0 \text{ or } 5$
75.	$\frac{-7+2\sqrt{5}}{5} \approx -0.5$ $\frac{-7-2\sqrt{5}}{5} \approx -2.3$
76.	a. $x^2-12x+36$ b. $x^2+\frac{1}{2}x+\frac{1}{16}$
77.	a. $(x-6)^2$ b. $\left(x+\frac{1}{4}\right)^2$
78.	No, the equation must equal "0".

79.	Yes, a = $\frac{1}{5}$, b = -3, and c = 0. However, it would be easier to use the Quadratic Formula if you first cleared the fraction in $\frac{1}{5}x^2$ by multiplying both sides of the equation by 5.
80.	First, multiply both sides of the equation by −5 to make it $x^2-15x=0$. $\frac{15\pm\sqrt{(-15)^2-4(1)(0)}}{2(1)}\to\frac{15\pm\sqrt{225}}{2}$ $\to\frac{15\pm15}{2}\to 15 \text{ or } 0$
81.	a. $x-2=\pm9\to x=11 \text{ or } -7$ average: $\frac{11+(-7)}{2}\to\frac{4}{2}=2$ b. $x+8=\pm\sqrt{7}\to x=-8\pm\sqrt{7}$ average: $\frac{-8+\sqrt{7}+(-8-\sqrt{7})}{2}\to\frac{-16}{2}=-8$
82.	$y=-2(-5)^2+15\to y=-2(25)+15\to y=-35$
83.	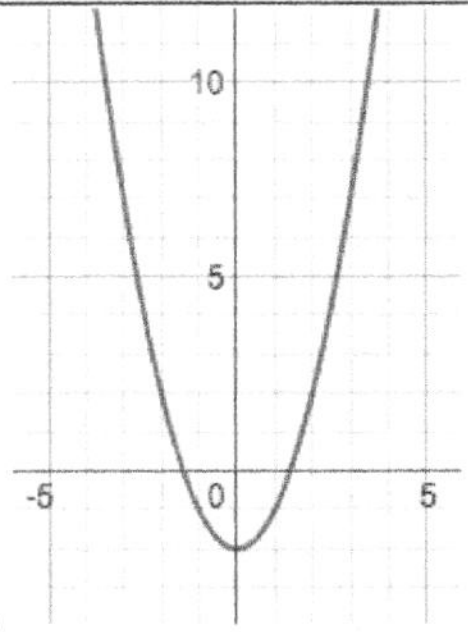 a. 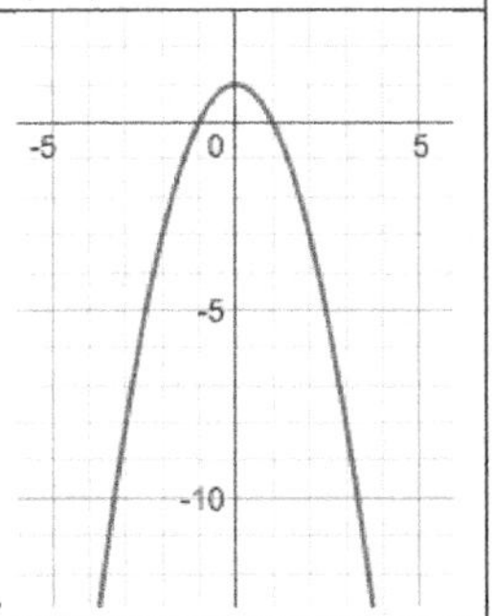 b.
84.	a. (0, −2) b. (0, 1)
85.	a. *x*-int: (0,0), (4,0); vertex: (2,2) b. *x*-int: (−6,0), (−2,0); vertex: (−4,−2)
86.	The *x*-value of the vertex is halfway between the *x*-intercepts
87.	(5, 0)
88.	*x* = 1.5 (halfway between -1 and 4)
89.	a. 2 b. 0.5 c. $\frac{x+5}{2}$ d. 4
90.	(8,0)
91.	(−1,0) and (11,0)
92.	a. 0 b. -2.5
93.	$3\to$ halfway between $3+\sqrt{6}$ and $3-\sqrt{6}$.
94.	a. 0 b. 9 c. $\frac{2}{7}$
95.	a. 7 b. *F* c. $\frac{y}{3}$ d. $\frac{-b}{2a}$

96.	a. $\frac{-b}{2a}$ b. $\frac{-b}{2a}$
97.	$\frac{-(-2)}{2(4)}\to\frac{2}{8}\to\frac{1}{4}$
98.	a. $\frac{-(-10)}{2(-1)}=\frac{10}{-2}=-5$ b. $\frac{-(-2)}{2\left(-\frac{1}{5}\right)}=2\div-\frac{2}{5}=2\cdot-\frac{5}{2}=-5$
99.	a. 0 b. −2.5 c. 3
100.	a. plug *x*-value into the function b. −16
101.	a. (3,12) b. (0,4) c. (2,−6)
102.	(−5,6)
103.	-
104.	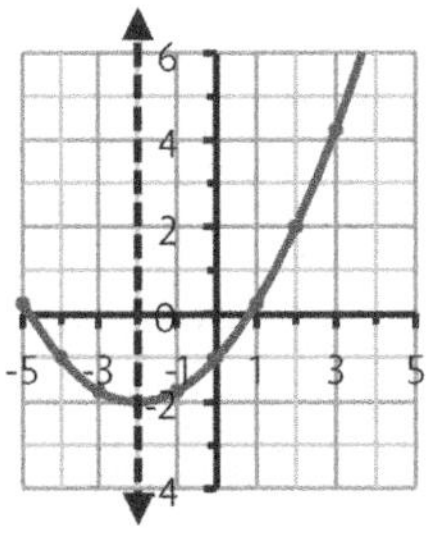 a. 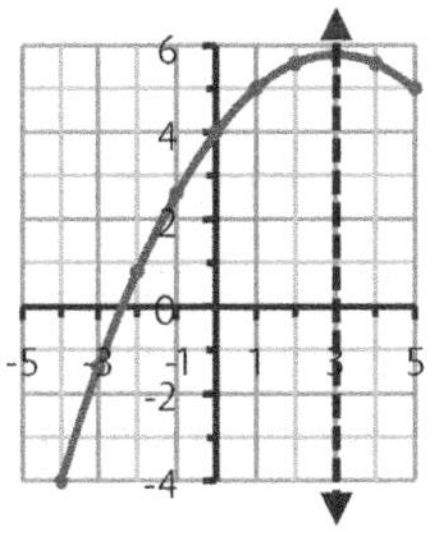 b.
105.	a. *x* = −2 b. *x* = 3
106.	*x* = −3; find the vertex: $x=\frac{-(-12)}{2(-2)}=-3$
107.	(0, −7); replace *x* with 0 and solve for *y*.
108.	(0, 11) ; replace *x* with 0 and solve for *y*.
109.	a. $x=20-2\cdot100\to x=-180$ b. $x-4=\pm3\to x=4\pm3\to x=7 \text{ or } 1$ c. $x(x-2)=0\to x=0 \text{ or } 2$
110.	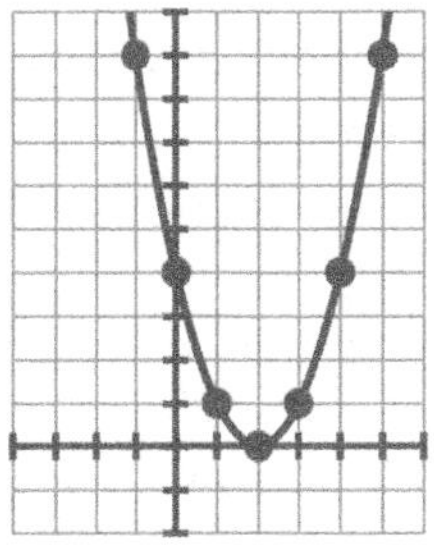

111.	
112.	
113.	
114.	vertex: (3,−1); no x-intercepts
115.	a. 1 b. 1.2
116.	In a quadratic function $y=ax^2+bx+c$, if the value of "a" is positive, the parabola opens upward. If the value of "a" is negative, the parabola opens downward.
117.	a. downward b. upward c. downward
118.	a. b. c.
119.	a. No: opens downward b. No: negative y-intercept c. No: no x-intercepts d. Yes e. No: one x-intercept is (0,0)

120.	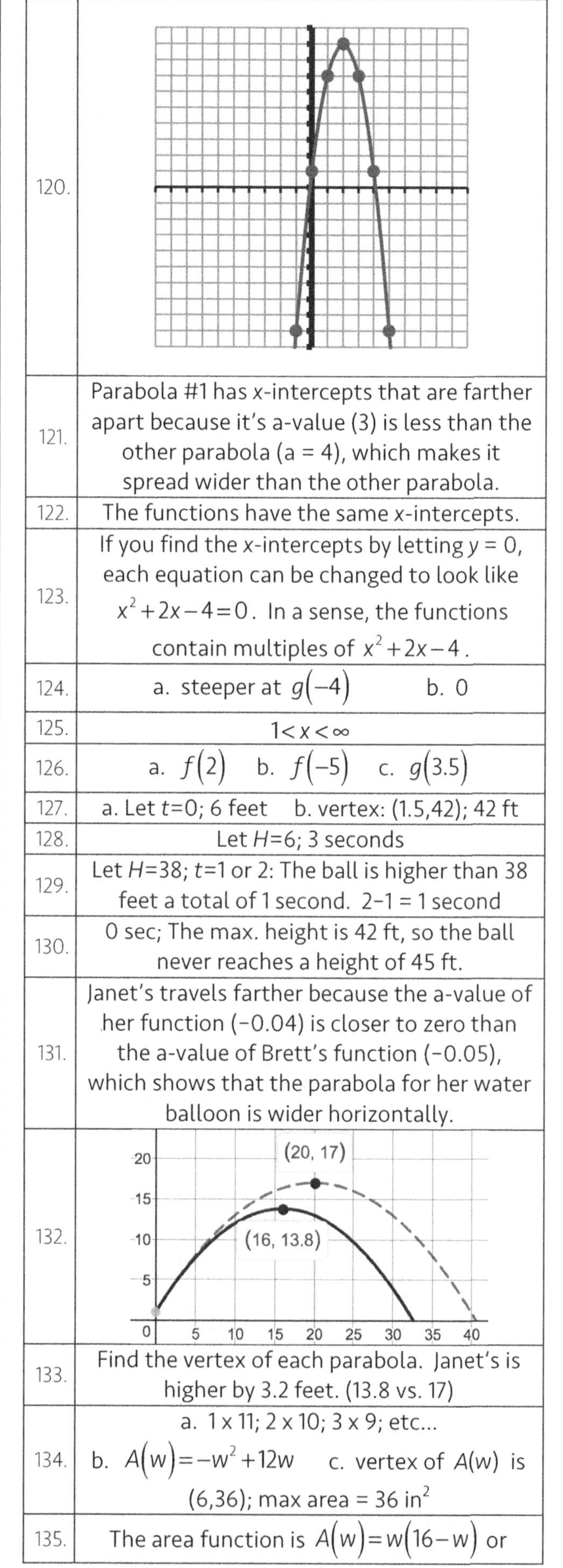
121.	Parabola #1 has x-intercepts that are farther apart because it's a-value (3) is less than the other parabola (a = 4), which makes it spread wider than the other parabola.
122.	The functions have the same x-intercepts.
123.	If you find the x-intercepts by letting y = 0, each equation can be changed to look like $x^2+2x-4=0$. In a sense, the functions contain multiples of x^2+2x-4.
124.	a. steeper at $g(-4)$ b. 0
125.	$1<x<\infty$
126.	a. $f(2)$ b. $f(-5)$ c. $g(3.5)$
127.	a. Let t=0; 6 feet b. vertex: (1.5,42); 42 ft
128.	Let H=6; 3 seconds
129.	Let H=38; t=1 or 2: The ball is higher than 38 feet a total of 1 second. 2−1 = 1 second
130.	0 sec; The max. height is 42 ft, so the ball never reaches a height of 45 ft.
131.	Janet's travels farther because the a-value of her function (−0.04) is closer to zero than the a-value of Brett's function (−0.05), which shows that the parabola for her water balloon is wider horizontally.
132.	(20, 17) (16, 13.8)
133.	Find the vertex of each parabola. Janet's is higher by 3.2 feet. (13.8 vs. 17)
134.	a. 1 x 11; 2 x 10; 3 x 9; etc... b. $A(w)=-w^2+12w$ c. vertex of $A(w)$ is (6,36); max area = 36 in^2
135.	The area function is $A(w)=w(16-w)$ or

	$A(w)=-w^2+16w$. The vertex of $A(w)$ is (8,64), so the max. area is 64 in^2.
136.	a. Let H=30; 20 feet b. When H=30, t=20 or 150: a total of 130 feet horizontally.
137.	55 yards is 165 feet. Calculate H(165), which is 8.25 feet. The ball is below the crossbar when it has traveled 165 feet (55 yards).
138.	a. Let H=10 and solve for x approx. 145.6 feet or 48.5 yards
139.	Yes. In the function, replace d with 56.7 to get H = 0.5362. Since H is positive, the shot put is still above the ground when it passes the national record distance.
140.	Her throw (57.1) breaks the record by about 0.4 feet, or about 5 inches.
141.	25.5 feet; find the x-value of the vertex of the parabola.
142.	a. No. The profits might rise at first, but eventually the profits would drop because she would not have enough customers. b. better profits at \$1.60 (\$20,000) than \$1.20 (\$18,000) c. monthly profits drop to \$14,000
143.	a. b.

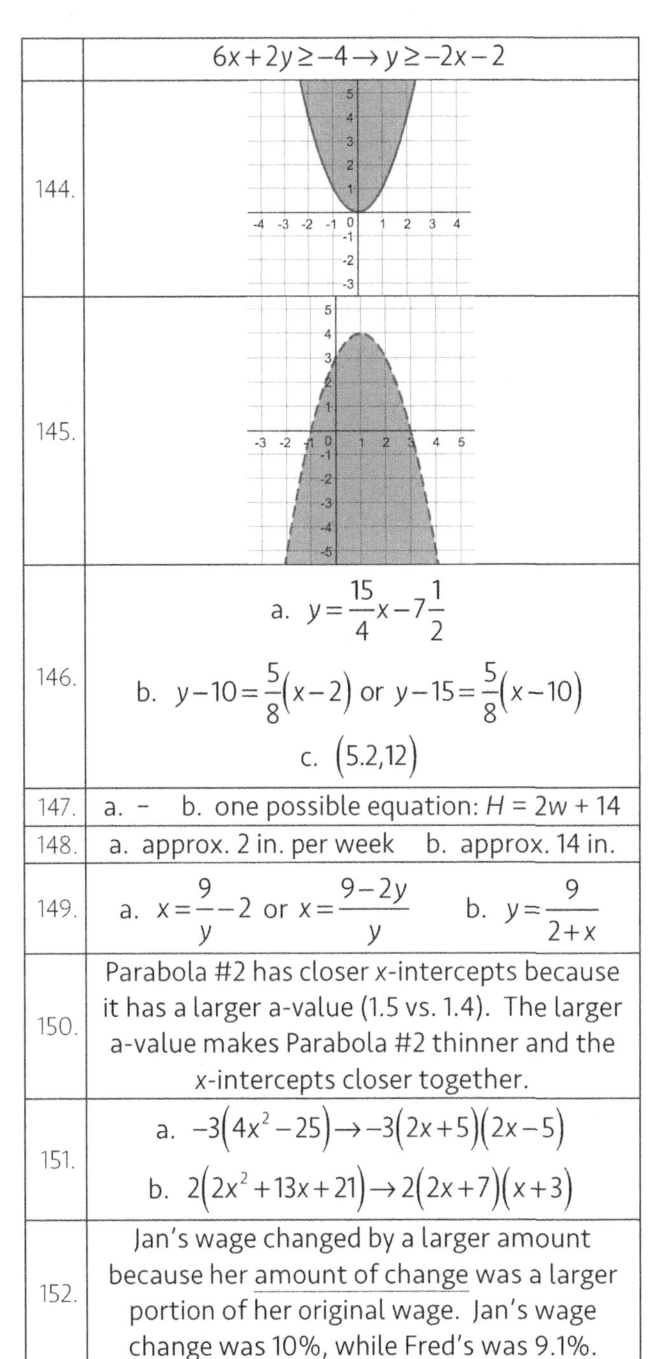

	$6x+2y\geq-4\rightarrow y\geq-2x-2$
144.	
145.	
146.	a. $y=\frac{15}{4}x-7\frac{1}{2}$ b. $y-10=\frac{5}{8}(x-2)$ or $y-15=\frac{5}{8}(x-10)$ c. $(5.2,12)$
147.	a. – b. one possible equation: H = 2w + 14
148.	a. approx. 2 in. per week b. approx. 14 in.
149.	a. $x=\frac{9}{y}-2$ or $x=\frac{9-2y}{y}$ b. $y=\frac{9}{2+x}$
150.	Parabola #2 has closer x-intercepts because it has a larger a-value (1.5 vs. 1.4). The larger a-value makes Parabola #2 thinner and the x-intercepts closer together.
151.	a. $-3(4x^2-25)\rightarrow-3(2x+5)(2x-5)$ b. $2(2x^2+13x+21)\rightarrow2(2x+7)(x+3)$
152.	Jan's wage changed by a larger amount because her amount of change was a larger portion of her original wage. Jan's wage change was 10%, while Fred's was 9.1%.